THE SILICON POWERPLAY

Power, Politics and the Global Chip Economy

Dr. Deepak Rao

Monalisa Bai

Copyright Page

The Silicon Powerplay - Power, Politics and the Global Chip Economy

ISBN: **978-81-938895-3-4** First Edition: 2026

Published by: UCCA. Printed in India

Disclaimer

Acknowledgements

This book is the result of curiosity, discipline, and a deep respect for the invisible systems that shape our world.

We would like to acknowledge the countless engineers, scientists, researchers, and industry pioneers whose work has built the semiconductor ecosystem over decades. Their contributions—often unseen—form the foundation of modern civilization.

We are grateful to the institutions, publications, and open sources of knowledge that made this work possible. Every effort has been made to appropriately credit and reference original sources wherever required. In the event that any reference has been unintentionally omitted, we sincerely regret the oversight and remain committed to correcting it in future editions.

We also acknowledge the limited use of artificial intelligence tools in supporting aspects of writing and refinement. These tools were used selectively, while the core ideas, analysis, and narrative remain the authors' original work.

Finally, we thank the readers. Your curiosity about the forces shaping the future is what gives this work meaning.

FOREWORD

Semiconductors have quietly become the foundation of the modern world. They power not only the digital experiences we interact with daily, but also the infrastructure, connectivity, and intelligence that define progress in the twenty-first century. What makes this industry truly distinctive is not just its technological complexity, but the way it sits at the intersection of innovation, economics, global collaboration, and long-term strategic thinking.

The Silicon Powerplay arrives at a defining moment and thoughtfully captures this intersection. It traces the evolution of the semiconductor industry from its scientific origins to its present role as a critical enabler of global innovation and economic resilience. As traditional linear progress models give way to system-level innovation, advanced packaging, heterogeneous integration, and deep cross-border interdependence, the industry can no longer be understood solely through scaling laws or manufacturing nodes. By blending technology, industry history, and high-level geopolitical context, the authors offer a holistic perspective on why semiconductors matter—not only to engineers and technologists, but to societies, economies, and policymakers worldwide.

In doing so, the book illustrates why semiconductors extend far beyond the boundaries of a conventional industry. They increasingly shape how nations think about capability, resilience, preparedness, and participation in the future digital economy. As supply chains become more specialized and geographically distributed, the semiconductor ecosystem reflects a broader reality of our time—where access, trust, continuity, and capability are as critical as invention itself.

One of the most important shifts underway is the convergence of design-led manufacturing and manufacturing-led design. Advanced semiconductors today cannot be realized through isolated excellence in either domain. Design choices depend on manufacturing realities—materials, packaging, thermal constraints, and yield economics—while manufacturing innovation is increasingly driven by architecture, software-hardware co-design, and application-aware system optimization. The future belongs to ecosystems that view design and manufacturing as a continuous, integrated loop rather than sequential handoffs. Within this evolving global landscape, India's semiconductor journey is both timely and consequential. India does not seek merely to participate at the margins, nor to replicate legacy models of the past. Its aspiration is to emerge as an equal and trusted partner in the global semiconductor ecosystem, contributing meaningfully across design, manufacturing, packaging, talent development, and system innovation—for India and for the world. From an India ecosystem standpoint, the timing of this book is particularly relevant, as the country stands at an important inflection point—building on its strength in design and engineering talent to develop deeper capabilities across the semiconductor value chain.

Semiconductors exemplify a fundamental truth of the modern era: no single country, company, or institution can build this ecosystem alone. A chip designed in one geography depends on tools, materials, software, manufacturing, packaging, and talent distributed across many others. This intrinsic interdependence, while introducing vulnerabilities, also forms the foundation for shared innovation. The semiconductor industry remains deeply specialized in its technologies, yet profoundly interconnected in its impact. What I particularly appreciate in this work is its balanced and forward-looking approach. Rather than framing the industry through narrow or adversarial lenses, it emphasizes collaboration, capability building, and long-term innovation. These are the principles that will shape sustainable semiconductor ecosystems worldwide. Enduring leadership in this field will belong not only to those who innovate boldly, but also to those who build trusted ecosystems, cultivate lasting partnerships, and recognize that progress emerges through both excellence and connection.

As the world looks toward a future shaped by artificial intelligence, climate transition, digital infrastructure, and intelligent systems at scale, semiconductors will remain the invisible enabler beneath visible transformation. How nations collaborate—on technology, talent, standards, and trust—will determine whether innovation advances fragmented interests or a better collective future. This book is a valuable contribution to that conversation, especially for readers seeking to understand how a technology hidden beneath layers of silicon and metal has become a defining force of our time—and how the decisions made today will influence the trajectory of innovation for decades to come.

— Navin Bishnoi
Vice President & India Country Manager, Marvell Technology
Chairperson, India Electronics & Semiconductor Association (IESA)

ABOUT THE AUTHORS

Dr. Hon. Major Deepak Rao is a polymath and multidisciplinary thinker whose work spans strategic studies, applied mind sciences, and human performance, integrating insights across medicine, Zen philosophy, military training, and political science. A renowned close-quarters battle (CQB) specialist and reflex shooting pioneer, he is also among the very few non-celebrity individuals to have been conferred an honorary rank by the President of India alongside prominent names such as Mahendra Singh Dhoni and Abhinav Bindra. With executive education from globally reputed institutions such as MIT, Harvard, Stanford and Wharton, he combines scientific reasoning with leadership and cognitive frameworks to explore decision-making under extreme conditions. His intellectual contributions include works like A Comprehensive Analysis of World Terrorism, referenced in United Nations–linked contexts, and Encyclopaedia of Close Combat Ops, which has found place in international security and research libraries, including those associated with agencies such as the FBI and Interpol. While widely known for his contributions to combat training, his broader body of work is distinctly multidisciplinary, reflecting a rare synthesis of tactical knowledge, philosophical depth, and analytical rigor, positioning him as a unique voice at the intersection of strategy, cognition, and global security.

He is the creator of the POWERPLAY Series, a body of work examining how power evolves across physical, technological, and cognitive domains. In his other complementary book, AI POWERPLAY: Command & Control in the Digital Battlespace, Deepak Rao advances this framework into the domain of artificial intelligence—

analyzing how influence, infrastructure, and cognition converge to reshape control itself.

Monalisa Bai is a semiconductor professional working at the intersection of technology, strategy, and AI-driven transformation. With over 15 years of experience across global environments involving complex manufacturing ecosystems, product lifecycles, and cross-functional execution. Her career reflects a rare blend of technical depth and strategic perspective, shaped by exposure to international fabs, evolving supply chains, and the increasing convergence of data, automation, and semiconductor innovation. Over time, she has expanded beyond execution into analyzing how semiconductors influence economic power and technological leadership, integrating real-world operational insights with broader strategic thinking on AI-enabled manufacturing and global supply chain resilience. In this book, she brings a multidisciplinary, forward-looking lens to the global silicon power play, bridging technical, business, and policy domains to offer a grounded yet strategic interpretation of how semiconductor ecosystems are reshaping the future of innovation, economic strength, and geopolitical power. Through this work, she seeks to contribute to a deeper understanding of how semiconductor ecosystems will shape the next phase of global power, innovation and technological sovereignty.

Table of Contents

INTRODUCTION

In every era, power has had a visible face—armies, oil, industry, and capital. Today, however, power has become increasingly invisible, embedded not in what we see, but in what we depend on without noticing. At the heart of this transformation lies the semiconductor chip: a creation small enough to rest on a fingertip, yet powerful enough to underpin the architecture of the modern world.

The Silicon Powerplay: Power, Politics and the Global Chip Economy explores this hidden force that drives modern civilization—and the global struggle to control it. It offers a compelling lens through which to understand shifting geopolitical balances, the fragility of interconnected systems, and the far-reaching consequences of technological innovation.

This book makes the invisible visible. It reveals how billions of microscopic transistors orchestrate the functioning of economies, militaries, healthcare systems, and digital life itself. From the transformation of ordinary sand into ultra-pure silicon to the creation of complex chips that power artificial intelligence and global networks, it traces the extraordinary journey of semiconductors—the unseen engines of the modern age.

Moving beyond technology, the narrative expands into a high-stakes geopolitical contest. It examines the rise of Silicon Valley, Japan's ascent and decline in semiconductors, Taiwan's strategic centrality, and the intensifying competition among the United States, China, Europe, and India. In this new world order, chips are no longer mere components—they are instruments of power. Control over semiconductor technology is no longer just an economic advantage; it is a strategic imperative.

The book also looks ahead, exploring the future of computing—from the slowing of Moore's Law to breakthroughs in artificial intelligence chips, advanced packaging, quantum computing, and photonic systems. As the industry approaches physical and economic limits, innovation is shifting from shrinking transistors to reimagining architectures, materials, and design itself.

Blending science, industry, and global strategy, The Silicon Powerplay reveals how a technology hidden beneath layers of metal and silicon has become the foundation of modern life—and a decisive force in shaping the future.

This is essential reading for anyone seeking to understand the forces defining the twenty-first century. It reminds us that the most consequential battles of our time may not be fought with weapons or words, but with wafers, code, and the quiet hum of machines.

PART I: The Invisible Technology Running the World

Chapter 1: The Chip That Runs Civilization – The Invisible Engine of Modern Life

Historically, the foundations of power were readily identifiable: oil-fueled empires, steel-built armies, and financial capital that influenced markets and governments. While these traditional sources of strength remain relevant in the twenty-first century, a more subtle and often overlooked force has emerged as increasingly decisive. Unlike the visible symbols of power, this force operates quietly within machines, concealed by glass, metal, and plastic. Nevertheless, it plays a critical role in determining global leadership, technological advancement, and interdependence among nations. This force is embodied in the semiconductor chip.

The reliance of modern civilization on minuscule slivers of silicon—semiconductor chips—underscores their critical importance. These engineered materials, small enough to rest on a fingertip, support the infrastructure of the contemporary world. They power smartphones and cloud servers, guide missiles and satellites, enable hospital equipment, operate banking systems, and facilitate artificial intelligence models that require immense computational resources. The disappearance of semiconductors from daily life would lead to a swift dismantling of the digital era. Communication channels would break down, financial infrastructures would grind to a halt, and logistics networks would cease to function. Furthermore, the operational capacity of modern vehicles, industrial facilities, and military systems—along with the broader stability of global economies—would be fundamentally compromised. What appears as convenience on the surface is, in reality, a deep structural dependence on silicon.

Semiconductor chips, though rarely visible to the average person, are integral to modern life. Typically concealed beneath heat spreaders, soldered onto circuit boards, and sealed within devices, they often go unnoticed unless a malfunction occurs. However, their invisibility does not equate to insignificance. Despite their small size, these chips embody highly sophisticated architectures, representing a convergence of logic, design, materials science, and advanced manufacturing. They are widely regarded as among the most intricate manufactured products.

The fundamental building block of every semiconductor is the transistor, a microscopic switch that regulates the flow of electrical current. Although the function of transistors—repeatedly turning on and off—appears simple, this binary operation underpins nearly all aspects of digital civilization. Computation, memory, communication, sensing, imaging, encryption, navigation, and machine learning all result from the coordinated activity of billions of transistors. Modern processors may contain tens of billions of these switches, densely packed at scales measured in nanometers, which are thousands of times smaller than a human hair. The precision required in chip manufacturing now resembles atomic-level choreography rather than traditional industrial processes. At nanometer scales, engineering is no longer just precision—it becomes a contest with the limits of physics itself.

Externally, a semiconductor wafer may appear unremarkable, resembling a polished, reflective disc. However, this appearance belies the extraordinary precision required in its manufacture—a process so exacting that only a select few companies globally can execute the most advanced steps. The fabrication process involves depositing, patterning, etching, and polishing multiple layers with exceptional accuracy. Photolithography enables the creation of features far smaller than intuitive understanding would suggest. Semiconductor fabrication facilities, or fabs, maintain rigorous standards of cleanliness and control, as even a single contaminant can compromise chips of significant value. In contrast to casual discussions of innovation, the semiconductor industry exemplifies the demanding, costly, and unforgiving nature of true technological advancement.

The significance of semiconductor chips extends well beyond the realm of electronics, as the industry underpins the entire digital economy. Telecommunications systems depend on semiconductors to route data globally. Transportation relies on them for navigation, engine control, safety mechanisms, and increasingly for autonomous driving. Defense applications utilize chips for radar, surveillance, secure communications, and precision targeting. Financial institutions require them to operate data centers, process transactions, and secure sensitive information. In healthcare, semiconductors are essential for scanners, monitors, implantable devices, and diagnostic tools. The entertainment sector, including gaming consoles and streaming infrastructure, is similarly dependent. Thus, the semiconductor chip has become the infrastructure underlying all other infrastructures.

This dynamic has created a powerful new reality: companies that drive semiconductor technology have become among the most influential institutions globally, even though their identities are often less recognized than those of consumer-facing brands. Intel, for example, pioneered the modern microprocessor and played a pivotal role in shaping the era of personal computing. Beyond manufacturing chips, Intel established foundational principles for distributing computing power to millions. For decades, its products were central to desktops, laptops, and servers, positioning Intel as a cornerstone of digital modernity.

Taiwan Semiconductor Manufacturing Company (TSMC) exemplifies the critical role of contract manufacturing in the global economy. TSMC is widely regarded as one of the world's most important manufacturers of advanced semiconductor chips. The company perfected a model that separates design from manufacturing, enabling specialized firms to focus on architecture while relying on TSMC to realize their designs. This division of labor transformed the semiconductor industry and introduced new concentrations of risk that are only now being fully recognized.

NVIDIA illustrates how technological shifts can alter the balance of power within the semiconductor industry. Initially recognized for its graphics processors in gaming, NVIDIA gained significant prominence when its hardware proved optimal for the parallel processing requirements of artificial intelligence. As machine learning models expanded in scale and complexity, NVIDIA's processors became central to many large-scale AI training and inference workloads, widely adopted in data centers. The company's market valuation reflected its perceived role as a driver of future technological advancement. Notably, hardware originally designed for rendering graphics has become instrumental in advancing artificial intelligence.

Companies such as Intel, TSMC, NVIDIA, Samsung, ASML, Applied Materials, Tokyo Electron, and others collectively constitute a highly complex global supply chain. No single nation or company controls all stages of semiconductor production. For example, a chip designed in California may rely on architecture teams in Austin, software tools from the United States, lithography equipment from the Netherlands, specialty chemicals from Japan, wafers from Taiwan, memory from South Korea, packaging in Malaysia or Vietnam, and final assembly elsewhere in Asia. This intricate network exemplifies globalization in tangible form—efficient, interdependent, and inherently fragile.

The vulnerability of the semiconductor supply chain became evident during the shortages that affected the global economy in the early 2020s, triggered and intensified by the COVID-19 pandemic, natural disasters, and geopolitical shocks, such as the Ukraine war. Automotive manufacturers reduced production due to insufficient chip supplies, and consumer electronics experienced significant delays. These disruptions highlighted the critical role of industrial supply chains in economic resilience. Modern vehicles, for instance, are complex systems reliant on numerous chips for functions such as braking, engine timing, entertainment, sensors, and safety. The semiconductor shortage demonstrated that even large corporations with extensive manufacturing capabilities can be immobilized by the absence of a single, small component. The semiconductor supply chain is not simply globalized—it is structurally distributed, with critical dependencies concentrated in specific segments.

Consequently, semiconductors have emerged as instruments of geopolitical power. Governments now regard advanced chips not merely as tradable goods, but as strategic assets. Access to state-of-the-art semiconductor technology influences economic productivity, military capability, intelligence operations, cyber power, and leadership in emerging technologies. Thus, access to advanced computation is becoming strategically significant, as access to energy was in earlier eras. Control over advanced chips and the tools required for their production confers significant leverage across multiple domains.

That realization has intensified international competition. The United States has taken steps to strengthen domestic semiconductor production and to tighten export controls on certain advanced technologies consistent with national security interests. China has invested heavily in building greater semiconductor self-sufficiency, driven by the recognition that technological dependence is a strategic vulnerability. Europe has sought a larger role in advanced manufacturing and design. South Korea and Japan continue to deepen their positions in memory, materials, tools, and manufacturing. India, historically stronger in software and design-related capabilities than in semiconductor fabrication, has begun carving out its own place in the ecosystem through incentives, partnerships, and industrial ambition. Many countries seek a larger place in the semiconductor ecosystem, but only a limited number currently command the full range of advanced capabilities.

Central to this competition is the reality that control over semiconductor capabilities increasingly determines influence over the future digital landscape. Nations that secure supply chains, talent, design expertise, manufacturing scale, and access to critical tools will shape both markets and strategic outcomes. Whereas access to sea routes or oil reserves once defined national power, in the current era, technological factors such as nanometer-scale manufacturing, foundry capacity, and advanced packaging are equally decisive.

However, the history of semiconductors encompasses more than competition and interdependence; it is also a testament to human ingenuity. The revolution began with the invention of the transistor, which replaced vacuum tubes and fundamentally transformed electronics. Vacuum tubes were large, fragile, inefficient, and generated excessive heat. In contrast, transistors offered reduced size, enhanced reliability, lower power consumption, and scalability. This breakthrough not only improved existing systems but also enabled the development of entirely new technologies.

The subsequent development of the integrated circuit marked another significant advancement. Engineers succeeded in placing multiple transistors on a single silicon substrate, creating compact circuits capable of increasingly complex functions. This innovation initiated the industrial-scale miniaturization of electronics, enabling modern computing. The integrated circuit established a new paradigm of progress, with each generation of chips offering greater functionality, enhanced performance, and broader accessibility.

Moore's Law encapsulated the rapid pace of progress in the semiconductor industry, positing that the number of transistors on a chip would approximately double every two years. Although not a physical law, it became a guiding principle that the industry actively pursued. Over several decades, chipmakers achieved remarkable gains in performance, cost efficiency, and density. Computers evolved from room-sized machines to desktops, laptops, and ultimately to portable devices. Technologies that once required significant institutional resources became widely accessible consumer electronics, fundamentally transforming both technology and society.

The smartphone exemplifies this transformation. Devices routinely used for communication, navigation, photography, banking, entertainment, and work now possess greater computing power than the systems that supported the Apollo missions. This comparison, though frequently cited, underscores the extraordinary pace of semiconductor advancement. Capabilities once reserved for governments, militaries, and research institutions are now accessible to individuals worldwide.

The semiconductor industry now faces significant challenges as the traditional model of progress—continually shrinking transistors—becomes increasingly difficult to maintain. As feature sizes approach atomic scales, physical limitations such as heat dissipation, power leakage, fabrication complexity, and escalating costs present formidable obstacles. Further miniaturization is no longer solely an engineering issue but also a confrontation with the fundamental constraints of physics and economics.

This shift does not signal the end of innovation, but rather a change in its direction. The industry is now exploring alternative architectures, novel integration methods, and new materials instead of relying solely on transistor miniaturization. Advanced packaging has emerged as a critical frontier, enabling the combination of multiple chiplets into integrated systems that offer enhanced performance and flexibility. This evolution alters both the economic and design paradigms of computing, allowing processors to be constructed as interconnected systems rather than as monolithic units.

This transition is evident throughout the semiconductor industry. AMD has implemented chiplet-based designs to enhance performance and scalability in both data center and consumer processors. TSMC and Intel are investing substantially in advanced packaging technologies that facilitate the integration of multiple dies. NVIDIA's artificial intelligence systems increasingly depend not only on processing power but also on the efficient assembly of processors, memory, and interconnects into comprehensive computational platforms. In the memory sector, high-bandwidth packaging is becoming essential for AI workloads. The industry's future will focus on optimizing the collaboration of multiple components rather than solely reducing the size of individual chips.

Researchers are also advancing beyond conventional silicon by exploring alternative materials. Compounds such as gallium nitride and silicon carbide are increasingly utilized in power electronics, electric vehicles, renewable energy systems, and fast-charging infrastructure due to their superior ability to manage higher voltages and temperatures in specific applications. While silicon remains central to the industry, it is now complemented by specialized materials tailored to particular functions. Sustained technological leadership often depends on adaptation rather than permanence.

Simultaneously, new computing paradigms are being developed. Quantum computing offers the potential for transformative capabilities in specific problem domains, although it is not yet positioned to replace conventional processors for routine tasks. Photonic computing investigates the use of light to process and transmit information more efficiently, while neuromorphic architectures seek to emulate aspects of brain structure to enhance certain computational processes. Although these fields are still in early stages and subject to considerable promotional attention, it is evident that the future of computation may involve multiple architectural approaches.

Artificial intelligence is also transforming the semiconductor industry internally. AI is increasingly used as a tool in chip design, helping engineers optimize layouts, improve verification processes, and accelerate development across various stages. Notably, while semiconductors have enabled the advancement of AI, AI is now contributing to the design of subsequent generations of semiconductors.

The evolution of semiconductors is not concluding with the limitations of silicon. Instead, the field is becoming increasingly competitive, innovative, and consequential. The forthcoming phase will be defined by advancements in manufacturing, national policy initiatives, supply chain resilience, materials science, software-hardware integration, and strategic competition among nations and corporations. The central issues now encompass not only technical challenges but also economic, political, and societal dimensions.

It is indisputable that semiconductors underpin the modern era. They define who leads it. They power devices, support economies, and increasingly shape the global balance of power. Positioned at the intersection of technological innovation, strategic interests, commerce, and policy, semiconductors are essential to understanding the dynamics of the twenty-first century.

The era of silicon remains central to contemporary technological development. Its influence on the future continues to intensify.

References

1. Bell Labs, 2022, The Transistor: 75 Years Since the Famed Bell Labs Invention Changed the World.
2. Nobel Prize, 1956, The Nobel Prize in Physics 1956.
3. Nobel Prize, 2000/2002, The Transistor in a Century of Electronics.
4. Intel, 2025, Moore's Law.
5. Semiconductor Industry Association, 2022, 2022 State of the Industry Report.
6. TSMC, 2024, Annual Report 2024.
7. NVIDIA, 2026, Form 10-K Annual Report.
8. Samsung Electronics, 2025, Business Report.
9. ASML, 2024, Annual Report 2024.
10. Applied Materials, 2025, Annual Report.
11. Reuters, 2021, Why Is There a Global Chip Shortage and Why Should You Care?
12. Reuters, 2022, Ukraine Halts Large Share of Neon Output for Chips.
13. Hall, E. C., 1996, Journey to the Moon: The History of the Apollo Guidance Computer.
14. UCIe Consortium, 2025, UCIe Specification.
15. NIST, 2022, CHIPS for America.
16. NIST, 2022, A Strategy for the CHIPS for America Fund.
17. European Commission, 2023, European Chips Act.
18. Ministry of Economy, Trade and Industry Japan, 2024, Outline of Semiconductor Revitalization Strategy in Japan.

19. Bureau of Industry and Security, U.S. Department of Commerce, 2023, Export Controls on Advanced Computing and Semiconductor Manufacturing Items.
20. Bureau of Industry and Security, U.S. Department of Commerce, 2024/2025, China-Related Semiconductor Export Control Updates.
21. India Semiconductor Mission, 2026, Mission Overview.
22. Nature Electronics, 2024, Advanced Packaging of Chiplets for Future Computing Needs.
23. Nature Electronics, 2024, High-Performance, Power-Efficient Three-Dimensional Heterogeneous Integration of Chiplets.
24. TSMC, 2024, Advanced Packaging and 3DFabric.
25. U.S. Department of Energy, 2026, Power Electronics R&D: SiC and GaN Wide-Bandgap Materials.
26. Nature Review Article, 2025, Wide-Bandgap Semiconductors and Power Electronics as Pathways to Carbon Neutrality.
27. IBM, 2025, IBM Quantum Roadmap.
28. Nature, 2024, Partial Coherence Enhances Parallelized Photonic Computing.
29. MIT News, 2024, Photonic Processor Could Enable Ultrafast AI Computations.

Chapter 2: From Sand to Silicon – How Matter Became Intelligence

The origin of every semiconductor chip is unexpectedly ordinary. Rather than beginning in a laboratory, a high-tech factory, or a clean room, the process starts with sand. The transformation that follows is not merely industrial—it is the systematic conversion of common matter into advanced capability.

Sand, a ubiquitous and often overlooked material, accumulates on beaches, deserts, and riverbeds, and has historically served as a resource for glass, construction, and other utilitarian purposes. However, within sand lies silicon, a fundamental component of modern technology. From this common substance originate computers, smartphones, satellites, cloud infrastructure, weapons systems, medical devices, and the processors that enable artificial intelligence. Despite the complexity of the digital world, its foundation remains this unremarkable material.

Silicon is the second most abundant element in the Earth's crust, exceeded only by oxygen. It typically occurs in compounds such as silicon dioxide, which constitutes much of sand and quartz, a crystalline form of silicon dioxide (silica). For centuries, silicon-containing materials were utilized in glassmaking, ceramics, and construction without recognition of their future technological significance. The transformation of this elemental foundation into a critical component of global infrastructure was unforeseen.

While raw sand is abundant, semiconductor manufacturing relies heavily on extremely high-purity silica or quartz, a significant portion of which is currently sourced from specific pegmatite deposits known for their low impurity profiles. Silicon for modern integrated circuits must reach an extremely high level of purity, often 99.999999999 percent purity, known as the 11N benchmark. At nanometer scales, even a minimal trace of impurities can significantly alter electrical properties, introduce defects, reduce performance, or compromise entire batches of chips. So, at this scale, purity is not a quality metric—it is a prerequisite for predictable behavior.

Carbothermic reduction is the initial transformation step in which the purified quartz and carbon are heated in a furnace at temperatures around 2,000 degrees Celsius, producing metallurgical-grade silicon. Although this material represents a significant improvement over raw quartz and is suitable for various industries, it remains insufficiently pure for semiconductor applications.

Silicon must be highly purified for chip manufacturing, so it undergoes several refining steps to remove impurities. Chemical processes turn silicon into gases or compounds such as trichlorosilane, which are then purified and redeposited in cleaner forms. What appears to be a chemical refinement is, in reality, the elimination of variability at a scale where even atomic-level inconsistencies matter. A key step is the fractional distillation of trichlorosilane, followed by deposition methods that produce ultra-pure polysilicon. By the conclusion of this process, the silicon attains a level of purity previously considered unattainable, which is essential for the reliability and commercial viability of modern electronics.

In addition to purity, semiconductor manufacturing requires silicon with a highly ordered atomic structure to ensure precise electrical behavior. This is achieved through single crystal growth, most commonly using the Czochralski method. In this process, purified polysilicon is melted in a crucible, and a seed crystal is introduced and slowly withdrawn while the crucible is rotated. As the seed is pulled upward, atoms align in a consistent crystalline pattern, forming a large cylindrical ingot. What is being engineered here is not just a material, but a controlled atomic order on which all subsequent precision depends.

The single-crystal growth stage requires precise control of temperature, rotation speed, and pull rate, as small variations can alter the crystal structure and introduce defects. The resulting ingots may weigh several hundred kilograms and reach diameters of 300 millimeters or more. Their primary value lies in their structural precision, as these crystals serve as the foundation for billions of transistors.

After the ingot is formed, it is sliced into thin circular discs called wafers. These wafers serve as the foundational surfaces for semiconductor fabrication. Engineers construct intricate layered structures on these wafers, which are essential for the production of processors, memory chips, and server-grade CPUs.

Prior to fabrication, the wafer must be polished to achieve near-atomic flatness and cleanliness. Microscopic irregularities are unacceptable, as subsequent circuit layers require precise alignment. Even minor scratches or contaminants can significantly reduce yield or performance. Consequently, wafer preparation and fabrication occur in ultra-clean environments with continuous air filtration, where contamination is rigorously controlled.

Once prepared, the wafer enters the fabrication plant, or fab, which represents one of the most advanced manufacturing environments globally. Leading-edge fabs require investments of tens of billions of dollars and integrate materials science, optics, robotics, chemistry, software, and precision engineering. These facilities operate as highly controlled ecosystems, where every variable is monitored, and failures can result in high costs.

Within the fab, the wafer undergoes hundreds of processing steps, including thin film deposition, etching, and doping with selected impurities to modify electrical properties. Layers are constructed and patterned with high precision. Each step must align accurately with preceding and subsequent processes, as the final device results from the cumulative effect of numerous precise interventions.

Lithography is a critical stage in semiconductor fabrication, involving the projection of nano-scale circuit patterns onto the wafer to define transistor and interconnect structures. For advanced chips, this process relies on highly specialized equipment, such as extreme ultraviolet (EUV) machines produced primarily by ASML in the Netherlands. These systems use very short-wavelength light to achieve feature sizes previously considered unattainable and rely on complex global supply chains for components such as mirrors, lasers, vacuum systems, precision stages, and software.

The global reliance on a limited number of advanced lithography machines highlights the semiconductor industry's concentration, fragility, and strategic importance. Delays in tool delivery, shortages of specialty materials, or geopolitical disputes involving key suppliers can create significant disruption across the entire digital economy. Although a fab may appear self-contained, it is actually part of a highly interdependent global network.

Through repeated lithography and fabrication steps - including thin-film deposition, ion implantation, etching, and chemical mechanical planarization - the wafer gradually becomes densely populated with billions of transistors, each functioning as a microscopic electrical switch. These Front-End-of-Line (FEOL) steps, supported by thermal oxidation and diffusion, create the microscopic electrical switches that form the fundamental active devices within the silicon substrate.

Once the transistors are formed, fabrication moves into the Back-End-of-Line (BEOL), where multiple layers of metallization and dielectric deposition build the intricate interconnect network that links the devices together. Throughout both FEOL and BEOL, continuous metrology and automated optical inspection are used to monitor precision, alignment, and structural integrity. When organized into carefully designed patterns, these interconnected structures become logic gates, memory cells, and processing cores capable of executing trillions of operations per second. This extraordinary combination of scalability, precision, and reliability stands among the defining achievements of semiconductor engineering.

After fabrication, the wafer contains numerous individual chips, which are separated through a process called dicing. Each resulting die represents a raw chip, though not every die will meet the performance and reliability standards. Rigorous testing is essential to ensure high yields, as economic viability depends on minimizing failures, regardless of design quality.

The functional dies that pass testing undergo packaging, during which they are enclosed and connected, and prepared for integration into larger systems. Advanced packaging is a critical area of innovation, influencing chip connectivity, heat dissipation, signal propagation, and the integration of multiple components. In applications such as artificial intelligence and high-performance computing, packaging provides significant competitive advantages.

Subsequently, chips are integrated into a wide range of devices, including laptops, smartphones, automobiles, servers, industrial robots, and telecommunications equipment. These chips facilitate medical diagnostics, navigation systems, financial transactions, and machine learning operations in data centers. The transformation from sand to silicon represents not only a material change but also a fundamental shift in technological capability.

The semiconductor manufacturing process relies on a highly specialized and interdependent global ecosystem. TSMC is widely recognized as a key manufacturer of many of the world's most advanced logic chips, while Intel has historically been a major integrated device manufacturer and an important contributor to semiconductor process and product development. Samsung is a major contributor to memory and advanced process technology. Equipment suppliers such as ASML, Applied Materials, and Lam Research provide essential fabrication machinery. Japanese companies supply critical chemicals, photoresists, and specialty materials, and packaging and assembly are frequently conducted in Southeast Asia. No single entity controls the entire process; each stage depends on expertise developed over decades and distributed internationally. Over time, capability in this domain is built through sustained coordination across materials, tools, processes, and talent.

The complexity of semiconductor manufacturing makes rapid replication challenging. While governments may set ambitious goals and investors provide capital, developing genuine capability requires time, skilled personnel, ecosystem maturity, and iterative learning. A fabrication facility represents accumulated expertise in yield improvement, process control, materials handling, tool calibration, defect management, and supplier coordination. The semiconductor industry exemplifies how modern power is rooted in deeply embedded technical competence as well as financial investment.

Over several decades, the semiconductor manufacturing process has undergone continuous improvement. Engineers have consistently increased transistor density, reduced costs per function, and expanded computing capabilities. Moore's Law broadly described the long-running trend toward increasing transistor density and computing capability over time. Achieving this progress required ongoing innovation in design, fabrication, materials, and equipment. As a result, computers have become smaller, more affordable, and more powerful, enabling the emergence of new industries.

The evolution of semiconductors is evident in the progression from early personal computer processors, which, though limited by current standards, transformed office work, education, and communication. Contemporary chips from companies such as Apple,

AMD, Intel, NVIDIA, and Qualcomm provide significant computational power in widely used devices. In data centers, accelerators from NVIDIA and others enable the training of large artificial intelligence models. The semiconductor content in automobiles has increased with electrification, connectivity, and software integration. Industrial equipment now relies on chips for control, automation, and optimization.

Despite the sophistication of modern technology, the process consistently begins with quartz, or silicon dioxide, a material often overlooked in daily life. Semiconductors exemplify both scientific advancement and the capacity to derive significant value from common materials.

The transformation of sand into silicon illustrates the convergence of chemistry, physics, engineering, and industrial disciplines to elevate a common material into advanced technological applications. Although the environments of a beach and a chip fabrication facility differ greatly, they are linked by a complex and remarkable supply chain.

A grain of quartz sand becomes purified silicon. Purified silicon becomes a crystal. The crystal becomes a wafer. The wafer becomes billions of transistors.

And those transistors, arranged with breathtaking precision, become the hidden engines of the modern world. From abundance to precision, this journey defines how modern technology is ultimately constructed.

References

1. Encyclopaedia Britannica, 2026, Silicon.
2. Siltronic, 2025, Annual Report 2024.
3. SUMCO, 2025, Sustainability Report 2025.

4. Materials Today Communications, 2021, Modeling and Application of Czochralski Silicon Single Crystal Growth.
5. Cambridge University Press, 2023, Integrated Circuit Fabrication.
6. MDPI, Engineering Proceedings, 2026, Electronic-Grade Silicon Wafer Production: Review and Update of Life Cycle Inventory (LCI) Data.
7. Siltronic, 2024, Products.
8. Applied Materials, 2026, Etch.
9. Applied Materials, 2026, Metals Deposition.
10. Lam Research, 2026, Etch Products.
11. Lam Research, 2024, Deposition Essentials.
12. ASML, 2026, EUV Lithography Systems.
13. ASML, 2024, 5 Things You Should Know About High NA EUV Lithography.

Chapter 3: The Transistor Revolution – The Tiny Switch that Changed Human History

Prior to the advent of the semiconductor era, electronic systems depended on vacuum tube technology, which is now considered largely obsolete.

The vacuum tube represented a significant technological advancement in its era, enabling the development of modern electronics during the first half of the twentieth century. Radios, radar systems, long-distance telephone networks, and early electronic computers all relied on this technology. Vacuum tubes controlled electrical signals by directing electrons through a vacuum between heated metal elements. While theoretically elegant, in practice vacuum tubes were cumbersome, fragile, energy-intensive, and generated substantial heat, often rendering advanced machinery inefficient and difficult to maintain. These constraints did not just limit performance—they defined the boundaries of what electronic systems could become.

Engineers accepted the limitations of vacuum tubes due to the absence of viable alternatives for amplification and switching in advanced electronic systems. However, these limitations significantly influenced system design. Machines were necessarily larger to accommodate the tubes, required extensive cooling to manage heat output, and demanded frequent maintenance due to high failure rates. Increased machine power necessitated more tubes, which in turn heightened the likelihood of operational failures.

The challenges associated with vacuum tubes became particularly evident in early computers. The Electronic Numerical Integrator and Computer (ENIAC), one of the first large-scale electronic computers, utilized nearly eighteen thousand vacuum tubes and weighed over thirty tons. Its operation required substantial physical space and electrical power. Maintaining functionality was a persistent challenge due to frequent breakdowns, excessive heat, and system complexity. A single tube failure could interrupt computations, making reliability a constant concern that required ongoing effort to achieve even temporarily. At this scale, computing was not yet a scalable technology—it was an engineered exception.

Such limitations characterized the pre-transistor era. Although electronic devices functioned, they were often inefficient, cumbersome, and unsuitable for widespread adoption. For computing to transition beyond laboratories, military facilities, or large institutions, a fundamentally different technological foundation was required. Engineers recognized this necessity, prompting the search for a viable alternative.

The search for a new foundation began to yield results at Bell Labs in the 1940s. Researchers at Bell Labs investigated the electrical properties of semiconductors, a class of materials that exhibit conductivity characteristics intermediate between metals and insulators. Unlike metals, which conduct electricity readily, and insulators, which resist it, semiconductors can be engineered to conduct or resist electrical flow in controlled ways. This property made them highly promising for signal control applications.

The ability to manipulate semiconductor conductivity fundamentally transformed electronics. Materials that could be engineered to either conduct or resist electrical flow provided a pathway to devices capable of replacing vacuum tubes without their associated size, fragility, or high-power consumption. Researchers at Bell Labs sought not only an improved component but also a new conceptual framework for electronic systems.

In 1947, John Bardeen and Walter Brattain demonstrated the first operational transistor at Bell Labs. Although modest in appearance, the device had profound implications. The transistor could amplify and switch electronic signals, fulfilling the essential functions of a vacuum tube in a significantly smaller and more efficient form. It required no heated filament, consumed less power, and generated less heat, indicating the potential for more compact and reliable electronic systems. The transistor did not simply replace the vacuum tube—it redefined the architecture of electronic possibility.

Shortly thereafter, William Shockley advanced transistor design with the junction transistor, which proved better suited to broader manufacturing adoption. This innovation established the foundation for modern semiconductor electronics. In recognition of their contributions, Bardeen, Brattain, and Shockley received the Nobel Prize in Physics, underscoring both the scientific and historical significance of their work. The transistor not only addressed an engineering challenge but also initiated a new technological era.

The advantages of transistors over vacuum tubes were substantial. Transistors were smaller, faster, more robust, and significantly more energy-efficient. They could be densely integrated, required less maintenance, and produced less heat. Crucially, transistors could be manufactured at industrial scale, marking a shift from bespoke assembly to mass production. The transistor thus represented not only an improved component but also the foundation of a new industrial paradigm for electronics.

The impact of transistors became evident during the 1950s as they replaced vacuum tubes in radios, military systems, telecommunications equipment, and early computers. The emergence of the transistor radio exemplified their commercial potential, enabling portable devices that consumers rapidly adopted. For the first time, advanced electronics became personal, portable, and integrated into daily life. The transistor radio signaled the onset of technological miniaturization.

Despite addressing many issues associated with vacuum tubes, early transistor-based systems introduced new challenges. Although transistors were smaller and more reliable, complex electronic systems still required assembly from numerous discrete components, often involving labor-intensive or inefficient processes. As system sophistication increased, this approach became increasingly inefficient and prone to errors. The primary limitation shifted from the performance of individual switches to the complexity of assembling large numbers of separate parts.

A significant advancement occurred when engineers recognized that multiple components could be integrated onto a single piece of semiconductor material, rather than constructing circuits from separate parts. This innovation revolutionized not only performance but also the fundamental structure of electronic manufacturing.

In 1958, Jack Kilby of Texas Instruments demonstrated the first integrated circuit, establishing that multiple electronic functions could be implemented on a single semiconductor substrate. Concurrently, Robert Noyce at Fairchild Semiconductor developed a more practical silicon-based manufacturing method for integrated circuits. While Kilby validated the concept, Noyce's contributions enabled scalable industrial production by using the planar process to create interconnected circuits on a single wafer, effectively laying the foundation for the modern semiconductor industry and highlighting the distinction between invention and manufacturability.

The advent of the integrated circuit marked the beginning of microelectronics. Engineers were no longer constrained to assembling individual components manually; instead, they could design and manufacture entire circuits as unified systems. This development significantly reduced size, cost, and power consumption while enhancing reliability. It also enabled a degree of complexity unattainable with previous models, allowing electronics to scale far beyond the limitations of vacuum tube-based systems.

The integrated circuit enabled the development of the microprocessor, fundamentally altering the relationship between individuals and computing technology. Computing power, previously confined to large institutional machines, could now be concentrated in compact forms and distributed broadly. This innovation served as the foundation for the personal computer revolution.

Several companies emerged to shape the evolving technological landscape. Intel, established in 1968, became a prominent figure in semiconductor history by producing commercially successful microprocessors that defined the computing era. Its products powered numerous desktops, laptops, and servers. Fairchild Semiconductor, although less prominent in later public memory, played a crucial role as an incubator of talent and innovation. Many of Silicon Valley's leading companies can trace their origins to individuals associated with Fairchild. Thus, the transistor revolution encompassed not only technological advancements but also the development of institutions, ecosystems, and new industrial cultures.

Advancements in chip technology enabled engineers to integrate increasing numbers of transistors onto a single chip. This progression was not solely for increased density; higher transistor counts translated into greater processing power, expanded memory, enhanced functionality, and reduced cost per operation. Consequently, each successive generation of chips offered improved capabilities within a smaller physical footprint.

In 1965, Gordon Moore noted that the number of transistors on a chip tended to double approximately every two years, an observation that became known as Moore's Law. While not a physical law, Moore's Law established an industry benchmark that drove continuous innovation. For decades, companies have advanced lithography, materials, architecture, and process control to sustain this rate of progress.

These advancements fundamentally transformed society. Computers that once occupied entire rooms were reduced to desktop size, then to laptops, and eventually to portable devices. Computational logic became embedded in appliances, industrial systems, vehicles, financial networks, and communication infrastructure, permeating nearly every aspect of modern life. The impact of the transistor extended across multiple sectors, as digital logic could be widely applied once affordable computation became available.

Illustrative examples from the semiconductor industry highlight the magnitude of this transformation. Early transistors enabled portable radios, while integrated circuits facilitated the development of calculators and compact military electronics. Microprocessor-powered personal computers produced by companies such as IBM and Apple have significantly contributed to the global PC industry's growth. Memory chips revolutionized data storage and retrieval. Mobile processors from companies like Qualcomm and Apple enabled the proliferation of smartphones. Graphics processing units (GPUs) from NVIDIA, initially designed for mostly graphics, have evolved into essential components for artificial intelligence and high-performance computing in today's era. Throughout these developments, the transistor has remained the foundational element, even as electronic systems have increased in complexity and capability.

The transistor revolution established the foundation for the digital economy. Global financial systems rely on processors to execute trades, clear payments, manage risk, and secure data. Communication networks depend on semiconductor chips for traffic routing, signal compression, and device connectivity. Cloud computing is supported by data centers equipped with processors and accelerators. Artificial intelligence training now requires significant semiconductor resources, including advanced GPUs and custom AI chips developed by leading technology firms. The transistor has evolved from a mere component to a critical enabler of modern commerce, infrastructure, and strategic operations.

However, transistor technology has continued to evolve. As engineers pursued further miniaturization, traditional designs encountered physical limitations. At nanometer scales, quantum effects become significant, leakage currents become increasingly difficult to control, and heat dissipation becomes more challenging. The assumptions underlying earlier scaling approaches have become less reliable due to these physical constraints.

To address these challenges, the industry has developed new transistor architectures. FinFETs, which feature fin-like structures extending from the chip surface, enhance current control and maintain performance as scaling becomes more complex. More recently, gate-all-around designs have been introduced, enclosing the channel with the gate structure to achieve even greater electrical control. These innovations demonstrate the industry's capacity to reimagine transistor design when traditional approaches reach their limits.

This ongoing evolution is reflected in the strategies of leading semiconductor manufacturers. Companies such as Intel, TSMC, and Samsung have made substantial investments in advanced transistor technologies to sustain performance and power efficiency. NVIDIA's AI accelerators utilize dense, high-performance transistor structures alongside advanced packaging techniques. Apple's custom silicon demonstrates how sophisticated transistor integration can yield significant consumer performance improvements. In all instances, the transistor remains the fundamental unit of technological advancement, despite increasing system complexity.

The transistor, initially developed as a small experimental device at Bell Labs, has become central to one of the most significant technological revolutions in history. It enabled electronics to become portable, scalable, affordable, and widespread. The transistor transformed computing from an exclusive industrial capability into a mass phenomenon, laying the groundwork for the digital world well before the concept of digital transformation became widely recognized.

Across a wide range of applications, from radios and rockets to calculators, data centers, mainframes, and smartphones, the transistor has remained the fundamental building block of modern electronic systems. Extensive software, network infrastructures, and artificial intelligence platforms are all built upon this essential component. The history of the transistor illustrates that significant societal change often results from incremental technical innovations that redefine the boundaries of possibility.

From a microscopic switch to a universal enabler, the transistor not only enhanced electronic technology but also transformed the scale at which computational intelligence could be integrated into machines.

References

1. Nokia Bell Labs, History.
2. Nokia Bell Labs, The Transistor: 75 Years Since the Famed Bell Labs Invention Changed the World.
3. Nokia Bell Labs, 1956 Nobel Prize in Physics.
4. Nobel Prize, The Nobel Prize in Physics 1956.

5. Nobel Prize, John Bardeen – Nobel Lecture: Semiconductor Research Leading to the Point Contact Transistor.
6. Computer History Museum, ENIAC.
7. Computer History Museum, Timeline of Computer History: ENIAC.
8. Smithsonian National Museum of American History, Integrated Circuit by Jack Kilby.
9. Texas Instruments, From Idea to Invention: The Origin Story of the Tiny Chip That Changed the World.
10. Computer History Museum, 1960: First Planar Integrated Circuit Is Fabricated.
11. Intel, Understanding Moore's Law.
12. Intel, Semiconductor Manufacturing Process.
13. Intel, Intel 18A.
14. IBM Research, VTFET: IBM's Revolutionary New Chip Architecture.
15. IBM Newsroom, IBM Research Alliance Builds New Transistor for 5nm Technology.
16. IBM Newsroom, IBM and Samsung Unveil Semiconductor Breakthrough That Defies Conventional Design.
17. Samsung Semiconductor, GAA Structure Transistors: A Next-Generation Process for Next-Generation Semiconductors.
18. Samsung Semiconductor, Samsung Opens the Gate to Transistor Performance, Power and Area Improvements with MBCFET.
19. TSMC, 2nm Technology.
20. Nature Electronics, 2018, Nanosheet FETs at 3 nm.
21. Nature Communications, 2024, Integrated 2D Multi-Fin Field-Effect Transistors.

Chapter 4: The Age of Integrated Circuits – When Complexity Learned to Shrink

The invention of the transistor addressed a major challenge in early electronics, yet significant limitations remained.

The replacement of vacuum tubes with transistors enabled machines to become smaller, more efficient, faster, and more reliable, marking a revolutionary advancement. However, as engineers sought to construct increasingly powerful systems, they encountered a new obstacle. Although less dramatic than the failures associated with vacuum tubes, this challenge was equally restrictive. Despite the use of transistors, electronic circuits still required assembly component by component. Each transistor needed to be connected to resistors, capacitors, wires, and other elements, often by hand or through labor-intensive manufacturing processes. As system complexity increased, both the number of components and the interconnections between them multiplied.

This issue escalated rapidly. While systems with a limited number of components remained manageable, those with hundreds became challenging, and those with thousands introduced overwhelming complexity, increased costs, and heightened risk of failure. Engineers referred to this escalating burden as the "tyranny of numbers," a term that accurately captured the situation. Although electronics had advanced to enable the conceptualization of highly complex systems, the practical task of wiring numerous components was becoming unmanageable. Even with flawless individual transistors, the vast number of solder joints and interconnections introduced significant opportunities for error. Thus, increasing complexity threatened the practicality of further advancements. At this stage, the constraint was no longer the device itself, but the complexity of connecting it into functioning systems.

The electronics industry required a fundamentally new approach rather than incremental improvements to the transistor. A paradigm shift was needed: circuits had to be conceptualized as unified structures rather than as assemblies of discrete components. This breakthrough occurred in 1958 and, as with many transformative innovations, initially attracted little attention.

In 1958, Jack Kilby at Texas Instruments demonstrated one of the earliest integrated circuits. This achievement was both conceptually elegant and highly consequential. Rather than assembling circuits from separate components, Kilby demonstrated that multiple electronic elements could be fabricated simultaneously on a single semiconductor wafer. By integrating circuits into a unified structure, the complexity of wiring and the vulnerability of numerous connections were significantly reduced, directly addressing the tyranny of numbers.

Concurrently, Robert Noyce at Fairchild Semiconductor developed a related yet more practical approach. Utilizing silicon and the planar process, Noyce helped advance a silicon-based planar approach that proved more suitable for scalable manufacturing. This distinction proved critical, as technological impact often depends not only on proof of concept but also on scalability. Noyce's contributions transformed the integrated circuit from an innovative concept into a viable industrial platform. The transition from invention to manufacturability marked the point at which technology began to scale meaningfully.

Collectively, the innovations of Kilby and Noyce initiated the integrated circuit revolution.

The impact of this development was substantial. Engineers could now fabricate entire circuits directly onto silicon, rather than assembling systems component by component. This transformation altered the economics, size, and reliability of electronic devices. Most significantly, it expanded the potential scope of electronic systems. The ability to construct circuits as single integrated structures enabled new approaches to scale, density, and repeatability.

The initial applications of integrated circuits emerged in contexts where reliability and compactness were paramount. Military systems were among the early adopters due to their need for robust electronics capable of functioning under demanding conditions. Space programs also adopted integrated circuits, as weight, power consumption, and component reliability were critical factors. In such environments, bulky and failure-prone electronics were unacceptable.

A notable early example occurred during the Apollo program, where NASA utilized integrated circuits in the Apollo Guidance Computer, the onboard system responsible for navigation during lunar missions. This application underscores the significance of the technology. The integrated circuit, originally developed to address wiring complexity, became integral to one of humanity's most prominent engineering achievements. This moment exemplifies the convergence of miniaturization and technological ambition. Miniaturization did not merely reduce size—it expanded the scope of what systems could achieve.

With advancements in manufacturing techniques, engineers recognized that integrated circuits offered scalability as well as convenience. Increasing numbers of transistors could be incorporated onto a single chip, enhancing functionality without a corresponding increase in physical size. This compounding improvement enabled the development of more sophisticated systems and generated demand for continued progress in integration density.

In 1965, Gordon Moore articulated this accelerating trend, observing that the number of transistors on integrated circuits appeared to double approximately every two years. This observation became known as Moore's Law. Although not a scientific law, Moore's Law served as both a description of industrial progress and a benchmark that the semiconductor industry pursued with considerable determination. What began as an observed trend evolved into a coordinated industry trajectory.

For decades, Moore's Law influenced the pace of technological advancement. Engineers continuously reduced transistor size, refined manufacturing processes, improved lithography, and enhanced material control. As a result, computers that once occupied entire rooms were miniaturized into desktop machines, portable devices, and eventually compact objects suitable for personal use. Computing power, once rare and expensive, became affordable, ubiquitous, and integrated into everyday life.

A significant advancement occurred in 1971 when Intel introduced the 4004 – widely regarded as one of the first commercially successful microprocessors. This development marked a pivotal moment in the evolution of integrated circuits. The central processing unit, previously constructed from numerous discrete components or multiple chips, could now be implemented on a single integrated circuit, consolidating the core computational functions into one chip.

The implications of this innovation were substantial. Engineers could design systems centered on a compact processor, eliminating the need to build processors from multiple components. This simplification reduced costs, improved reliability, and accelerated technological innovation. The microprocessor represented not merely an improved component, but a fundamentally new computing architecture.

Subsequently, microprocessors were rapidly adopted in calculators, control systems, instrumentation, industrial machinery, and ultimately personal computers. By the late 1970s and early 1980s, the personal computer revolution had begun. Intel emerged as a leading processor designer, while companies such as IBM, Apple, and Commodore developed systems that democratized computing across offices, homes, and educational institutions. The integrated circuit transitioned from a technical innovation to the driving force behind widespread technological transformation.

Over time, the degree of integration increased. Engineers expanded their focus beyond the central processing unit alone, integrating memory, input/output functions, communication modules, graphics capabilities, and specialized accelerators. The initial goal of reducing wiring complexity evolved into a comprehensive philosophy: maximize functionality on silicon, minimize inter-functional barriers, and enhance performance while reducing size and cost.

This transformation affected nearly every industry influenced by electronics. Telecommunications transitioned from analog infrastructure to digital networks based on integrated electronics. Consumer devices became smaller, more powerful, and more affordable. Industrial automation advanced as factories implemented chips for sensing, control, and process management. Medical technology progressed through the adoption of compact electronics in imaging, monitoring, diagnostics, and implants. The development of the internet relied on semiconductor advancements, including servers, routers, modems, and switching hardware.

The integrated circuit enabled the emergence of the modern software era. Advanced software systems require robust underlying hardware. Operating systems, databases, digital design tools, enterprise platforms, streaming services, search engines, cloud infrastructure, and machine learning systems all rely on highly sophisticated integrated circuits. Silicon not only supported software development but also facilitated its scalability.

Numerous examples from the semiconductor industry illustrate this progression. Intel's early processors were instrumental in defining the personal computing era. Memory chips produced by companies such as Micron, Samsung, and SK Hynix enabled large-scale data storage and rapid retrieval. Qualcomm's mobile processors facilitated the transformation of phones into handheld computers. Apple's custom silicon demonstrated the performance and efficiency gains achievable through tightly integrated design. NVIDIA's GPUs, as advanced integrated circuits, became essential for artificial intelligence workloads and data center computing. Each technological generation reinforced the principle that increased integration yields greater capability. Integration consistently shifts the boundary between what is complex and what becomes routine.

Integrated circuits are now deeply embedded in most modern electronic systems, affecting nearly all aspects of modern life. They are present in smartphones, laptops, electric vehicles, MRI machines, aircraft systems, industrial robots, payment terminals, routers, satellites, and large-scale data centers. Integrated circuits power cloud computing, digital entertainment, logistics platforms, social media, advanced weapons systems, and artificial intelligence. Billions are manufactured annually, executing instructions at speeds and scales beyond human comprehension.

However, the integrated circuit, like all technological advancements, faces significant challenges. As transistors have been reduced to nanometer scales, new difficulties have emerged. Heat management has become more complex, power leakage has become more problematic, and manufacturing defects have become more impactful. Quantum effects increasingly influence areas previously governed by classical engineering. The traditional approach of continual miniaturization is no longer sufficient.

Consequently, the next phase of integrated circuit development increasingly relies on novel manufacturing techniques and innovative design philosophies. Advanced lithography, particularly with the adoption of ASML's extreme ultraviolet systems, has become essential. Foundries such as TSMC and Samsung have advanced their process technologies significantly, while Intel has continued investing heavily to strengthen its manufacturing position. Advanced packaging techniques now enable the combination of multiple chiplets and specialized dies into highly capable systems. The future of integration often involves assembling multiple optimized components into a cohesive platform, rather than solely increasing the density of a single monolithic chip.

These developments are particularly significant in the context of artificial intelligence and high-performance computing. Training advanced AI models typically requires chips with extremely high transistor density, sophisticated memory architectures, high-speed interconnects, and efficient power management. Data-center processors, GPUs, AI accelerators, and networking chips exemplify both the legacy and the latest evolution of integrated circuits. While the initial objective was to integrate several components onto a single silicon substrate, the current goal is to coordinate billions of transistors, multiple functional blocks, and often multiple dies within systems designed to support next-generation digital infrastructure.

The integrated circuit originated as a solution to a practical engineering challenge: excessive components, wiring, and potential points of failure. However, it soon surpassed its initial purpose, becoming the foundation for modern computing, communications, automation, and digital services. The integrated circuit has fundamentally transformed how society operates, travels, heals, communicates, learns, and interacts with the world.

From the Apollo Guidance Computer to contemporary smartphones and from the first microprocessor to AI accelerators in cloud servers, integrated circuits have shaped the architecture of the modern era.

What began as a solution to the tyranny of numbers evolved into the technology that provides modern civilization with its speed, scale, and intelligence.

References

1. Texas Instruments, 2022, From Idea to Invention: The Origin Story of the Tiny Chip That Changed the World.
2. Computer History Museum, 1958, Successful Test of the First Integrated Circuit.
3. Computer History Museum, 1959, Practical Monolithic Integrated Circuit Concept Patented.
4. Computer History Museum, 1960, First Planar Integrated Circuit Is Fabricate
5. Intel, 1971, The First Programmable Microprocessor: The 4004.
6. Moore, G. E., 1965, Cramming More Components onto Integrated Circuits.
7. Intel, 1971, Announcing a New Era of Integrated Electronics.
8. Smithsonian National Air and Space Museum, 2015, Apollo Guidance Computer and the First Silicon Chips.
9. Computer History Museum, 2019, Silicon Chips Take Man to the Moon.
10. NASA, 2000, Computers Take Flight: A History of NASA's Pioneering Digital Fly-By-Wire Project.
11. ASML, 2024, EUV Lithography Systems.
12. ASML, 2024, 5 Things You Should Know About High NA EUV Lithography.

13. ASML, 2024, ASML and imec Open Joint High NA EUV Lithography Lab.
14. TSMC, 2024, Advanced Packaging Services.
15. TSMC, 2025, CoWoS®.
16. Hennessy, J. L., and Patterson, D. A., 2018, A New Golden Age for Computer Architecture.
17. Lau, J. H., 2021, Semiconductor Advanced Packaging.
18. IEEE Engineering and Technology History Wiki, 2015, First-Hand: Invention of an Integrated Circuit.

PART II: The Global Semiconductor Power Game

Chapter 5: The Rise of Silicon Valley – From Orchards to Global Power

By the middle of the twentieth century, the semiconductor revolution had already begun to gather force. The transistor broke the vacuum tube's dominance. Integrated circuits were opening the door to smaller, faster, and more reliable electronics. Yet the industry itself still lacked a true center of gravity. Talent was dispersed. Companies were young. The business's rules had not yet fully formed.

Subsequently, a single region began to emerge as the focal point of technological advancement.

That place was Silicon Valley.

Today, the name Silicon Valley appears inevitable, suggesting Northern California was always destined to become the center of the modern technology industry. However, prior to its association with semiconductors, software, venture capital, and technological ambition, the region south of San Francisco was primarily known for fruit production. Orchards dominated the landscape, with apricots, plums, and cherries thriving. The area was referred to as "the Valley of Heart's Delight," a name that evokes pastoral imagery rather than technological innovation.

But history has a habit of transforming ordinary landscapes into strategic terrain. In Silicon Valley, that transformation was driven not by geography alone, but by a combustible mix of science, entrepreneurship, ambition, and institutional support. The region's rise was not the work of a single inventor or company. It was the result of a chain reaction, and one of the key figures in that reaction was William Shockley.

Shockley had already secured his place in technological history as one of the inventors of the transistor at Bell Labs. Brilliant, ambitious, and undeniably consequential, he was also notoriously difficult. In the mid-1950s, after leaving Bell Labs, he moved to California and founded Shockley Semiconductor Laboratory in Mountain View in 1956. His goal was bold but entirely in character: build the next generation of semiconductor devices and create a world-leading center of electronic research.

The raw ingredients were there. Shockley had prestige. He had technical vision. He attracted highly talented young engineers. On paper, the company should have become a foundation stone of the new semiconductor age. Yet in technology, outcomes are shaped as much by organizational dynamics as by technical vision.

What began as an ambitious effort under Shockley Semiconductor Laboratory became an important early chapter in the emergence of Silicon Valley. Although the company did not evolve into the lasting industry platform that might once have been imagined, it brought together a remarkable group of talented engineers at a formative moment in semiconductor history. Historical accounts suggest that differences in leadership style, strategic direction, and working culture contributed to the departure of several key team members. In 1957, eight of these engineers left to establish Fairchild Semiconductor, a company that would go on to play a foundational role in the modern semiconductor industry. Later remembered in industry history as the “Traitorous Eight,” their departure is now widely seen not simply as a rupture but as a pivotal turning point that catalyzed the rise of Silicon Valley as a global center of innovation.

Fairchild quickly became one of the most important firms in the early semiconductor industry. It was not merely another startup. It was a laboratory of industrial transformation. At Fairchild, engineers developed manufacturing methods that enabled integrated circuits to become practical, scalable, and commercially viable. This was where the semiconductor industry began to mature from scientific promise into an industrial force. What emerged was not just a company, but a replicable model for innovation-driven growth.

Fairchild's importance went far beyond its products. It became a training ground, an incubator, and in many ways the ancestral source code of Silicon Valley itself. Some of the brightest engineering minds of the era worked there—ideas circulated quickly, and talent clustered. Ambition multiplied. The company generated not only chips but also people who would go on to find other companies, spread technical knowledge, and create the culture of entrepreneurial spinouts that became one of Silicon Valley's defining features.

Among the standout figures at Fairchild were Robert Noyce and Gordon Moore, two men whose names would become inseparable from the next phase of semiconductor history. Noyce helped develop the planar integrated circuit, one of the key breakthroughs that made chip manufacturing practical. Moore, sharp-minded and strategically influential, would later give the industry one of its most enduring formulations in Moore's Law. Together, they embodied a rare combination of technical depth and entrepreneurial instinct.

In 1968, Noyce and Moore left Fairchild and founded Intel.

At the time, Intel was among the promising entrants in a still-young industry. It began with a focus on memory chips, and early success in semiconductor memory gave it a foothold in the market. But Intel's real historical significance would emerge when it turned toward microprocessors. In 1971, the company introduced the Intel 4004 – widely regarded as one of the world's first commercially successful microprocessors. The achievement was enormous. A central processing unit (CPU) that once required a board full of components can now be placed on a single integrated circuit.

This development fundamentally transformed the field of computing.

The microprocessor did not merely improve electronics. It redefined the architecture of computing systems. Instead of designing a computer as a sprawling collection of custom-built logic, engineers could build machines around a compact, programmable processor. This reduced costs and simplified design opened the door to entirely new product categories. Calculators, industrial controllers, embedded systems, and eventually personal computers all emerged from this shift.

With Intel and other semiconductor firms gaining momentum, the region around them began to change. Engineers were drawn to the area. Entrepreneurs saw an opportunity. Investors started to recognize that technology firms, though risky, could also deliver outsized returns. A new kind of ecosystem began to take shape, one in which talent, capital, and technical ambition fed one another in a reinforcing loop. Over time, this feedback system became one of the region's most enduring advantages.

This systemic integration represented one of Silicon Valley's primary advantages. The region functioned as an interconnected ecosystem rather than a mere collection of companies.

Engineers moved between firms, carrying knowledge with them. Founders launched new ventures after leaving established companies. Investors were willing to fund ambitious technical ideas long before conventional business logic would have approved them. Stanford University played a critical role by producing skilled engineers, supporting research, and maintaining ties with the surrounding industrial community. The region became a place where ideas did not stay trapped inside institutions. They spilled outward, mixed, and returned in altered form.

Such circulation of knowledge proved to be significant. While many industries tend to restrict information, Silicon Valley fostered an environment where knowledge is frequently disseminated, recombined, and generated commercial innovation. The network effects that would later characterize internet platforms initially manifested through interpersonal relationships, frequent collaboration, overlapping expertise, and a culture that accepted failure as part of the innovation process. This dynamic arose not from inherent superiority, but from a structure that incentivized experimentation over caution.

By the 1970s and 1980s, the term "Silicon Valley" had taken hold. The name reflected the material at the heart of semiconductor manufacturing, but it also captured something deeper. Silicon was no longer just a chemical element. It had become the symbol of a new economic order. The region housed chipmakers, computer firms, and eventually software startups, all building atop the momentum created by semiconductor innovation.

Intel remained central to this story, but it was far from alone. Apple emerged and helped redefine personal computing by making computers more accessible, more appealing, and eventually more integrated into everyday life. The personal computer ceased to be a machine for specialists and became a tool for households, offices, students, and creators. Later, firms such as NVIDIA would demonstrate that semiconductor innovation could reinvent itself once more. Originally associated with graphics processing, NVIDIA's chips would become indispensable in fields ranging from gaming to scientific computing to artificial intelligence. The valley proved capable not only of launching industries, but of reinventing them.

Even when software and internet companies rose to dominate headlines, semiconductors remained the hidden core of the ecosystem. The internet boom of the 1990s depended on processors, memory, networking chips, storage systems, and server hardware. The software revolution did not float above the physical world. It sat squarely atop semiconductor progress. Silicon Valley's mythology often celebrated apps, founders, and venture rounds, but beneath the branding and bravado, the chip remained king.

The valley's culture also helped explain its durability. Risk-taking was normalized there in a way that would have seemed reckless in more traditional business environments. Startups were founded with incomplete products, ambitious claims, and the kind of confidence that often sits one step away from delusion. Most failed. A few succeeded spectacularly. Venture capital emerged as the financial engine of this system, supplying money to founders betting on uncertain futures. This changed the speed of innovation. Technical ideas no longer had to wait for slow-moving corporate approval or government sponsorship. If investors believed that the upside was large enough, capital could move quickly. This compressed the time between idea and execution in ways few other regions could replicate.

This model became one of Silicon Valley's most influential contributions to global innovation. The region fostered a distinctive approach: rapid development, talent acquisition, aggressive scaling, and the pursuit of market indispensability. Although this process was often inefficient and unpredictable, it proved highly effective.

By the early twenty-first century, Silicon Valley had become one of the world's most influential technology hubs. Its companies had reshaped computing, communications, commerce, entertainment, and increasingly geopolitics. But while the valley remained dominant in innovation, the semiconductor industry itself was changing.

Chip manufacturing was becoming more complex, more capital-intensive, and more specialized. The days when many companies could both design and manufacture cutting-edge chips were fading. Advanced fabrication plants require substantial investment, increasingly sophisticated process control, and access to a global supply chain of tools, chemicals, and materials. Building a leading-edge fab was no longer a matter of being smart and ambitious. It required billions of dollars, deep technical expertise, and an industrial ecosystem of extraordinary sophistication.

This pressure led to one of the most important structural changes in the semiconductor industry: the separation of design from manufacturing. Instead of doing both under one roof, many companies began focusing on chip design while outsourcing production to specialized foundries. This gave rise to the fabless semiconductor model.

The fabless approach changed the economics of the industry. A company could concentrate on architecture, product strategy, and customer relationships without carrying the crushing burden of operating its own manufacturing plants. This opened the door for a new generation of firms that could innovate rapidly without first spending fortunes on fabrication infrastructure. NVIDIA, Qualcomm, Broadcom, AMD (in its later form), and many others would thrive under this model. The separation of design and manufacturing redefined how value is created and captured in the semiconductor industry.

But the rise of fabless companies also elevated the importance of foundries, and no company would shape that world more than TSMC.

Taiwan Semiconductor Manufacturing Company transformed the industry by proving that contract chip manufacturing could become not merely viable, but dominant. Its rise shifted the center of gravity in chip production away from Silicon Valley and toward Asia, especially Taiwan. Silicon Valley remained a powerhouse of design, entrepreneurship, and technology strategy, but the physical fabrication of the world's most advanced chips increasingly took place elsewhere.

That shift changed the meaning of semiconductor power. The story was no longer just about inventors in California garages or startups along Sand Hill Road. It became a global story of supply chains, geopolitics, manufacturing concentration, and strategic vulnerability. Design might happen in California. Fabrication might happen in Taiwan. Equipment might come from Europe. Materials might come from Japan. Assembly might happen in Southeast Asia. The industry that Silicon Valley helped launch had outgrown its birthplace.

And yet the valley's role remains foundational. It created the culture, institutions, and company lineages that turned semiconductors into a world-changing industry. Shockley's dysfunctional lab gave rise to Fairchild. Fairchild gave rise to Intel and a generation of spinouts. Intel helped launch the microprocessor era. That era helped create the personal computer revolution, the software economy, the internet, and much of the technological landscape that followed.

Silicon Valley did not invent every part of the modern semiconductor world. But it created the environment in which many of the most important chapters could begin.

The region's transformation from agricultural land to a center of technological innovation illustrates the profound impact of its industrial evolution. The initial departure of eight engineers from Shockley's laboratory ultimately contributed to the creation of an ecosystem that has significantly influenced global society.

And by the time the world fully grasped what Silicon Valley had become, the semiconductor race it helped ignite had already gone global.

References

1. Nokia Bell Labs, History.
2. Encyclopaedia Britannica, 2026, William B. Shockley.

3. Computer History Museum, 1956, Silicon Comes to Silicon Valley.
4. Computer History Museum, Fairchild Semiconductor Founders.
5. Computer History Museum, 2016, Fairchild, Fairchildren, and the Family Tree of Silicon Valley.
6. Computer History Museum, 2017, Fairchild Semiconductor: The 60th Anniversary of a Silicon Valley Legend.
7. Stanford University, 2004, The Story of the Stanford Industrial Research Park.
8. Stanford Engineering, The Terman Era.
9. Encyclopaedia Britannica, Silicon Valley.
10. Oxford University Press, 2005, The Man Behind the Microchip: Robert Noyce and the Invention of Silicon Valley.
11. Macmillan, 2006, Broken Genius: The Rise and Fall of William Shockley, Creator of the Electronic Age.
12. Massachusetts Institute of Technology Press, 2006, Making Silicon Valley: Innovation and the Growth of High Tech.
13. W.W. Norton & company, 1997, Crystal Fire: The Invention of the Transistor and the Birth of the Information Age.
14. Intel, 1968, Intel's Founding.
15. Intel, 1971, Announcing a New Era of Integrated Electronics.
16. Intel, 1965, Moore's Law.
17. Encyclopaedia Britannica, Apple Inc.
18. NVIDIA, Our History: Innovations Over the Years.
19. NVIDIA, 2026, NVIDIA in Brief.
20. TSMC, 2024, Annual Report 2024.
21. TSMC, Company Profile.

Chapter 6: Japan's Lost Semiconductor Empire – From Global Leadership to Strategic Reinvention

By the late 1970s and early 1980s, the semiconductor industry had become a global phenomenon rather than solely an American narrative. While the United States retained its reputation for innovation, pioneering firms, and the mythology of Silicon Valley, Japan was quietly constructing a formidable semiconductor sector. Japan was assembling a semiconductor empire. This ascent reflected not only technical capability but the systematic application of manufacturing discipline at scale.

This transformation was deliberate. Japan had already established a global reputation for manufacturing excellence. Japanese companies revolutionized consumer electronics through rigorous attention to quality, efficiency, and precision. Products such as televisions, radios, and cassette players bearing Japanese brands became synonymous with reliability worldwide. The same approach that had succeeded in consumer electronics was now being applied to semiconductors.

The disciplines that strengthened Japanese electronics manufacturing also facilitated success in semiconductor production. Precision, yield, and process control were critical. Semiconductor manufacturing, characterized by strict tolerances and a reliance on consistency, provided an environment in which Japanese firms excelled. Companies such as NEC, Toshiba, Hitachi, and Fujitsu invested heavily in semiconductor production and established themselves as major industry players, ranking among the leading global semiconductor firms during what is often described as the 'Hinomaru Semiconductor' era.

Japanese firms' primary strength was in memory chips, particularly DRAM (Dynamic Random Access Memory). Although DRAM lacks the prestige of processors, it is essential for storing data and instructions required for computer operation. As personal computing expanded, demand for memory increased significantly. Japanese companies capitalized on this opportunity through disciplined, large-scale production. In this phase, competitive advantage was defined less by invention and more by execution.

Japanese semiconductor companies achieved success through manufacturing excellence rather than marketing strategies. They became highly proficient at producing large volumes of reliable chips with stringent quality control and operational efficiency. Investments in production capacity, process refinement, and yield improvement were prioritized. In an industry where minor defects can significantly impact profitability, these factors were crucial. By the mid-1980s, Japanese companies controlled a substantial share of the global memory chip market, at one point holding a dominant share of the global memory market, and a very large share of semiconductor production by some contemporary measures.

This level of dominance caused significant concern among United States industry leaders and policymakers.

Although American firms played a foundational role in the early development of the semiconductor industry, Japan's success demonstrated that long-term leadership depends not only on invention, but also on excellence in manufacturing, scale, and consistency. The ability to create a technology and the ability to produce it with exceptional efficiency are distinct strengths, and Japanese companies proved highly effective in the latter. This placed competitive pressure on firms such as Intel, which had been strong in memory chips, as expanding supply and changing market dynamics reduced margins across the segment.

For Intel, this became a defining strategic moment. The company chose to redirect its focus from memory toward microprocessors, aligning itself with a part of the industry that offered greater long-term opportunity. What may once have appeared to be a difficult transition ultimately became one of the most consequential strategic decisions in semiconductor history. By concentrating on microprocessors, Intel positioned itself at the center of the personal computing era and helped shape the next great phase of the digital age.

This strategic shift illustrates a key principle in the semiconductor industry: survival often favors companies that recognize when to adapt or exit specific markets, rather than those that persist without adjustment.

In the United States, Japan's rise came to be seen as more than a commercial development; it underscored the growing policy significance of semiconductors. American policymakers and industry leaders increasingly viewed Japan's progress through the broader lens of industrial competitiveness, market access, and national economic strategy. As a result, the discussion extended beyond corporate competition and became part of a larger policy conversation about the future structure of the global semiconductor industry.

The semiconductor industry had already become too economically and technologically significant to be treated as an ordinary sector, even before chips became a central focus of public policy and national industrial strategy.

These developments contributed to the 1986 U.S.-Japan Semiconductor Agreement, an important milestone that reflected the growing recognition of semiconductors as a matter of national policy as well as corporate competition. The agreement represented an effort to create a more structured framework for addressing concerns about market access and pricing practices, while also highlighting the industry's rising strategic importance. Although it emerged from a period of trade friction, it also marked the beginning of a broader transformation in the global semiconductor landscape. Japan had achieved extraordinary success, and the industry was entering a new phase in which the foundations of leadership were beginning to evolve.

Japan had achieved remarkable success, but the factors underpinning its dominance were beginning to shift.

In the 1990s, multiple factors began to undermine Japan's position. One significant factor was the shift in demand toward microprocessors and logic chips, driven by the rise of personal computers. American firms, particularly Intel, held advantages in these areas. Although memory chips remained important, they were no longer the sole focus. As the computing industry matured, architectural control and processing power became increasingly central to value creation.

Rising capital intensity also posed a significant challenge. Semiconductor manufacturing became increasingly technologically complex and expensive. Maintaining a leading position required ongoing investment in new fabrication facilities, advanced tools, improved process technologies, and highly skilled engineering talent. The industry entered a phase where only firms capable of absorbing substantial capital expenditures could remain competitive, placing pressure on all major participants, including Japan's leading companies.

A deeper structural transformation then began to reshape the industry. New competitors emerged in Asia, most notably with the development of Taiwan's foundry model. Rather than maintaining vertically integrated operations, Taiwanese firms specialized in manufacturing chips designed by other companies. This approach became one of the most significant business model innovations in semiconductor history.

TSMC, founded in 1987, was central to this shift. The company helped institutionalize the separation of chip design from manufacturing through the foundry model, enabling firms to participate in the semiconductor industry without the prohibitive cost of owning and operating expensive fabrication plants. Companies could concentrate on design, architecture, and product innovation while outsourcing manufacturing to specialized foundries.

This separation of design and manufacturing fundamentally transformed the structure of the semiconductor ecosystem, transitioning it from a landscape of closed, integrated giants to a highly specialised interdependent global supply chain.

As manufacturing became more capital-intensive, the fabless model gained appeal. Firms could allocate resources to designing advanced processors, communications chips, graphics chips, and custom accelerators without incurring the substantial costs of owning fabrication facilities. TSMC and other foundries assumed responsibility for manufacturing, thereby reshaping industry economics. The center of value began shifting from manufacturing ownership to design specialization and ecosystem coordination. Success in the subsequent era increasingly favored agile design firms partnered with leading foundries, rather than vertically integrated corporations.

Japanese semiconductor firms were not well-positioned for this transition. Many continued to focus on memory and maintained vertically integrated business structures. While these models had been effective in earlier industry phases, they became less adaptable as the balance of value shifted. The market increasingly rewarded flexibility, specialization, and cross-border collaboration. Although Japan's strengths in disciplined manufacturing persisted, the industry landscape was changing.

Consequently, Japan's share of the global semiconductor market declined sharply from its late-1980s peak to a much smaller level by the 2020s. However, this decline should not be interpreted as Japan's disappearance from the semiconductor industry. Japan remains highly important in several critical segments of the semiconductor value chain. Japanese companies lead in the production of key materials, including photoresists, silicon wafers, specialty chemicals, and other inputs vital to advanced chipmaking. Japan also plays a significant role in semiconductor equipment and precision manufacturing technologies. While Japan lost dominance in finished chip production, it retained substantial influence within the deeper layers of the supply chain.

This distinction is significant. Influence in the semiconductor industry extends beyond those who design CPUs or operate foundries to include suppliers of chemicals, materials, wafers, and machinery essential to the ecosystem. In this context, Japan's role evolved rather than disappeared, shifting from visible dominance in chip output to critical control over foundational elements of the semiconductor supply chain.

Nevertheless, Japan's experience offers an important lesson: leadership in the semiconductor industry is inherently transient. A country may achieve global prominence in one era but lose its position if it fails to adapt to technological advancements, evolving demand, and changing industrial structures. While precision and manufacturing excellence propelled Japan to the forefront, the industry continued to evolve toward new architectures, business models, and centers of influence.

This dynamic underscore the competitive logic of the semiconductor industry: it rewards innovation and excellence, but only temporarily, and continually demands reinvention.

By the end of the twentieth century, the center of semiconductor manufacturing was shifting once more, this time toward Taiwan. The Taiwanese model proved better aligned with the economic realities of the modern chip industry. Although Japan's semiconductor sector did not vanish abruptly, its period of dominance had concluded.

Japan's ascent demonstrated how rapidly manufacturing excellence can alter the global balance of power. Japan's subsequent decline illustrated the volatility of this balance. In the semiconductor industry, dominance is always contingent.

Japan's semiconductor sector experience is often interpreted as a reminder that industrial leadership can encourage assumptions of continuity that later prove fragile. Such advantages are rarely permanent. Thus, leadership is not sustained by past success, but by continuous adaptation to shifting technological and industrial realities.

References

1. NEC, 2020, 120 Years, and the Road Ahead.
2. NEC, 2025, History: Corporate Profile.
3. NEC, 1986, 1986 Memory Data Book.
4. Toshiba, 2019, Chronology of History.
5. Japan Semiconductor Corporation, 2016, Company History.
6. Negrine, J., Armstrong, S., and Drysdale, P., 2025, Chips in Japan: Industrial Policy, Decline and Renewal.
7. Dallmeyer, D. G., 1989, The United States-Japan Semiconductor Accord of 1986: The Shortcomings of High-Tech Protectionism.
8. WTO GATT, 1988, Japan – Trade in Semi-Conductors, Report of the Panel.
9. GATT, 1986, Arrangement Between the Government of Japan and the Government of the United States of America Concerning Trade in Semiconductor Products.
10. Intel, 1985, Farewell to DRAM.
11. Intel Alumni Network, 2022, Intel Memory Systems, an Anthology.
12. TSMC, 2025, About TSMC.
13. Ministry of Economy, Trade and Industry Japan, 2024, Outline of Semiconductor Revitalization Strategy in Japan.

14. CSIS, 2023, Japan Seeks to Revitalize Its Semiconductor Industry.
15. RIETI, 2025, "Chips in Japan: Industrial Policy, Decline and Renewal.
16. U.S. International Trade Administration, 2025, Japan, Semiconductors.
17. SUMCO, 2025, High Quality Silicon Wafers.
18. SUMCO, 2025, Management Policies.
19. AMRO, 2025, Japan's Strategic Comeback in the Global Chip Race.
20. Kaufman, C. S., 1994, U.S.-Japan Semiconductor Agreement: Chipping Away at Japanese Market Barriers, *UCLA Pacific Basin Law Journal*.
21. Johnson, C., 1982, MITI and the Japanese Miracle.
22. Callon, S., 1995, Divided Sun: MITI and the Breakdown of Japanese High-Tech Industrial Policy.
23. Fransman, M., 1990, The Market and Beyond: Cooperation and Competition in Information Technology Development in the Japanese System.

Chapter 7: Taiwan – The World's Most Important Island

Within the modern semiconductor industry, a single location has attained such significance that its stability can influence the direction of the global economy. This location is neither a vast continental power, a major military empire, nor a nation endowed with abundant oil or mineral resources. Rather, it is a relatively small island situated off the coast of China.

This island is Taiwan.

Despite its modest geographic size, Taiwan plays a disproportionately significant role in the structure of the modern world. It is central to the semiconductor supply chain, holding an exceptional level of strategic importance. Smartphones, laptops, servers, networking systems, data centers, military electronics, and artificial intelligence infrastructure all rely, directly or indirectly, on semiconductors produced in Taiwan. As computational power becomes increasingly vital, Taiwan has emerged as one of the most strategically significant locations globally.

Central to Taiwan's prominence is the Taiwan Semiconductor Manufacturing Company (TSMC).

Understanding Taiwan's significance requires examining the transformation initiated by TSMC. In the early decades of the semiconductor industry, most companies operated under the integrated device manufacturer (IDM) model, designing chips, fabricating them in-house, and selling the final products. While this structure was effective in earlier periods, it imposed substantial costs. Semiconductor manufacturing was inherently expensive, and costs escalated with each technological generation. Companies seeking to remain competitive were compelled to invest continually in new fabrication facilities, equipment, process technologies, and engineering expertise, resulting in an ever-increasing capital burden.

For many firms, sustaining the IDM model became increasingly challenging. Chip design and advanced manufacturing required distinct and increasingly specialized skill sets. It was no longer evident that successful chip designers needed to maintain their own fabrication facilities. The industry was therefore poised for structural transformation.

This transformation was driven by Morris Chang.

Chang is recognized as one of the most influential figures in semiconductor history, despite receiving less popular recognition than other prominent technology leaders. Having spent much of his career in the United States, including leadership roles at Texas Instruments, he possessed a deep understanding of both the technical and economic forces shaping the industry. When invited by Taiwan's government to develop a domestic semiconductor sector, Chang proposed not merely the creation of another chip company, but the adoption of an entirely new business model.

This concept was innovative in its simplicity. Rather than designing and manufacturing its own chips, a company could specialize exclusively in manufacturing chips for other firms. Designers could develop architectures, processors, memory controllers, or specialized accelerators without the need to construct their own fabrication facilities. Manufacturing responsibilities would be delegated to a dedicated foundry, whose primary function was to produce chips at scale with exceptional precision.

This represented one of the most significant shifts in the history of modern technology.

Many semiconductor startups possessed talented engineers, innovative concepts, and promising designs, but lacked the substantial capital required to construct and operate advanced fabrication facilities. The foundry model alleviated this financial burden, enabling firms to concentrate on design, product strategy, software integration, and customer requirements, while manufacturing specialists managed the complexities of production. This specialization facilitated a new era of semiconductor innovation.

Founded in 1987, TSMC institutionalized the dedicated foundry model and demonstrated its viability on a global scale.

The success of this model produced transformative effects. Companies that might have remained small design firms or not existed at all were able to become major semiconductor players. The fabless model emerged, enabling firms to focus on design while outsourcing production. Companies such as NVIDIA, Qualcomm, AMD, Broadcom, and MediaTek expanded within this ecosystem. TSMC effectively became the industrial foundation supporting a generation of semiconductor innovation. While these firms attracted public attention for their chip designs, TSMC served as the manufacturing engine that translated architectural concepts into physical products.

This transformation affected not only corporate structures but also the geographic distribution of the industry.

While Silicon Valley maintained its leadership in design, software, venture capital, and technology strategy, Taiwan became a central hub of advanced semiconductor manufacturing. TSMC constructed fabrication plants of exceptional complexity and scale, advancing process technology continuously. These facilities were among the most sophisticated manufacturing environments globally, with wafers undergoing hundreds of steps under stringent cleanliness and control. Even microscopic contaminants could damage circuits, and minor variations could impact yield. Precision was essential for operational success.

The economic challenges were similarly formidable. Constructing and equipping a leading-edge fabrication facility could require investments of tens of billions of dollars. The machinery, particularly advanced lithography systems, represented the pinnacle of industrial capability. The expertise necessary to operate these facilities effectively could not be acquired instantly; it had to be developed over years through engineering discipline, process optimization, and a culture focused on operational excellence.

This accumulated expertise became Taiwan's most valuable semiconductor asset.

TSMC continued to invest heavily, advancing from mature manufacturing processes to the most advanced technology nodes globally. As feature sizes decreased and transistor densities increased, the company became indispensable to customers requiring high-performance solutions. Smartphones demanded more powerful and efficient processors, data centers required higher-performance chips for cloud computing, AI companies needed accelerators for substantial computational workloads, and automotive systems increasingly relied on sophisticated semiconductors. In each case, the solution frequently originated in Taiwan.

As of 2025-2026, many of the world's most advanced chips utilized in flagship smartphones, high-performance computing systems, and AI infrastructure are manufactured by TSMC. The company's central role is such that the modern digital economy would likely face major disruption without its contributions. This situation reflects not only market share but also a high degree of industrial concentration, with much of the world's advanced chipmaking capacity located within a single ecosystem.

This concentration has significant implications.

Given the reliance of numerous global firms on TSMC, any significant disruption to Taiwan's semiconductor industry would likely have major ripple effects across the global economy. Such concentration transforms operational dependence into systemic vulnerability. Consumer electronics production would be impacted, automotive manufacturing could slow, cloud infrastructure expansion might stall, and AI development could encounter substantial bottlenecks. Already complex supply chains would experience further instability. Disruptions originating in Taiwan could propagate through factories, stock markets, defense industries, and data centers worldwide.

This dynamic explains Taiwan's sensitive position in global geopolitics.

Taiwan operates in a highly sensitive geopolitical environment, particularly in relation to China and the United States. As strategic competition among major powers intensifies, semiconductors have evolved from commercial products to assets directly linked to national security, technological leadership, military capability, and long-term economic resilience. Access to advanced chips influences leadership in artificial intelligence, communications, surveillance, autonomous systems, and defense technologies. Consequently, semiconductor manufacturing capacity is not only economically valuable but also geopolitically significant.

Taiwan is a focal point in this geopolitical tension.

Taiwan's significance derives not from geographic size or military power, but from industrial concentration and technical expertise. The island is considered indispensable to the semiconductor sector because it performs functions at a level few other locations can match. In an era frequently characterized by disruption, Taiwan's semiconductor sector demonstrates that certain forms of stability are exceptionally difficult to restore once lost.

This context has led analysts to adopt the term "silicon shield," suggesting that Taiwan's central role in semiconductor production provides a strong incentive for the international community to maintain regional stability. If the island is essential to the global technology system, major economies have a vested interest in preventing significant disruptions. While this concept is compelling, it does not guarantee security. The strategic importance of the supply chain may serve as a deterrent, but it does not eliminate risk. Historically, significant disruptions have occurred despite widespread awareness of potential consequences.

Nevertheless, the concept of the silicon shield reflects a genuine dynamic. Taiwan's increasing importance in the semiconductor sector has altered the perspectives of policymakers, investors, and strategists. The island is viewed not only through traditional diplomatic or security frameworks, but also as a foundational element of the modern computing infrastructure.

This recognition has prompted a global response.

Governments worldwide are making substantial investments in domestic semiconductor capacity. The United States has implemented significant policies to expand chip manufacturing domestically and reduce dependence on concentrated overseas production. Europe is working to rebuild and enhance its own semiconductor capabilities. Japan is strengthening its position through materials, equipment, and strategic partnerships. South Korea remains competitive in memory and advanced process technology, while India seeks to establish a presence in fabrication and packaging. The strategic importance of semiconductors has become widely recognized. The response reflects a broader realization that semiconductor capacity underpins long-term technological resilience.

Despite these global initiatives, Taiwan remains one of the central hubs of the advanced semiconductor supply chain.

This persistence stems from the fact that semiconductor leadership cannot be achieved rapidly. It demands more than financial investment; it requires a deep ecosystem, robust supplier relationships, a strong engineering culture, a skilled workforce, institutional memory, and a consistent record of overcoming complex manufacturing challenges. Taiwan has spent decades developing this foundation. Its expertise is embedded not only within companies, but also in individuals, routines, training systems, and industrial practices. Replicating this depth is significantly more challenging than implementing subsidy programs or making public policy announcements.

Taiwan's emergence as one of the global leaders is among the most notable developments in the modern global economy. This centrality was not accidental; it resulted from a combination of government support, industrial strategy, technical expertise, and disciplined execution. Taiwan transformed specialization into strategic leverage and manufacturing excellence into geopolitical significance.

The broader implication is significant. Historically, strategic power was often derived from territory, military capacity, or natural resources. In the digital era, it increasingly stems from mastery of a limited set of highly complex industrial processes. Taiwan exemplifies this shift more clearly than most other locations worldwide.

The semiconductor industry has redefined the concept of global significance. A small island became strategically central not through geographic dominance, but through technological mastery in semiconductor manufacturing.

As nations compete to secure access to advanced semiconductors, Taiwan occupies a central position in a new technological rivalry that will influence the trajectory of the twenty-first century.

In the modern semiconductor landscape, Taiwan exemplifies how concentrated expertise can translate into disproportionate global influence.

References

1. TSMC, About TSMC.
2. TSMC, Everything to Know About Dedicated Foundries.
3. TSMC, TSMC Arizona.
4. SEMI, Oral History Interview: Morris Chang.
5. Chang, M., 2021, The Founding of TSMC.
6. Nature Electronics, 2021, Taiwan and the Global Electronics Industry.
7. Council on Foreign Relations, 2026, U.S.-Taiwan Trade Agreement Leaves Major Questions Open.
8. Technological Forecasting and Social Change, 2006, Agile Strategy Adaptation in Semiconductor Wafer Foundries.
9. Research Policy, 2006, Explaining Taiwan's Semiconductor Industry Development.
10. Technological Forecasting and Social Change, 2003, The IC Fabless Industry in Taiwan: Current Status and Future Development.
11. Journal of Information and Optimization Sciences, 2002, Virtual Factory and Relationship Marketing: A Case Study of TSMC.
12. CSIS, 2025, Silicon Island: Assessing Taiwan's Importance to U.S. Economic Growth and Security.
13. CSIS, 2022, Taiwan's Semiconductor Dominance: Implications for Cross-Strait Relations.
14. CSIS, James A. Lewis Semiconductor Policy Work.
15. NIST, CHIPS News and Releases.
16. NIST, Research and Development Programs.
17. European Commission, European Chips Act.

18. Energy Research and Social Science, 2025, From Vulnerabilities to Resilience: Taiwan's Semiconductor Supply Chain.
19. The China Quarterly, 2021, Taiwan's Semiconductor Industry, the U.S.-China Tech Competition, and Cross-Strait Relations.
20. Semiconductor Industry Association and Boston Consulting Group, 2021, Strengthening the Global Semiconductor Supply Chain in an Uncertain Era.
21. U.S. Department of Commerce, 2021, Review of Critical Supply Chains.
22. National Science and Technology Council Taiwan, 2023, Science and Technology White Paper 2023–2026.
23. SEMI, World Fab Forecast Reports.
24. Brown, C., and Linden, G., 2011, Chips and Change: How Crisis Reshapes the Semiconductor Industry.

Chapter 8: The US-China Chip War – The Contest for Technological Leadership

In the twenty-first century, semiconductors transitioned from being mere components to serving as instruments of strategic leverage.

For decades, semiconductors were primarily regarded as commercial products, engineering milestones, or the concealed mechanisms within consumer devices. Their significance was largely articulated in terms of business and technological innovation. This perspective has shifted. Today, semiconductors occupy a central position at the intersection of economic power, military capability, artificial intelligence, industrial policy, and national security. They have evolved from being the engines of the digital world to being widely regarded as among the most strategic assets globally.

This transformation is most evident in the intensifying rivalry between the United States and China.

What may have previously appeared as a trade dispute or a technological competition has evolved into a broader contest over the future configuration of global power. Semiconductors are central to this contest. Advanced chips are indispensable for artificial intelligence systems, high-performance computing, cloud infrastructure, military applications, telecommunications networks, surveillance platforms, and emerging technologies such as autonomous vehicles and robotics. Control over advanced semiconductor capabilities increasingly shapes the pace and trajectory of innovation.

Throughout much of the modern semiconductor era, the United States held a leading position across several of the industry's most influential segments. American firms led in chip architecture, advanced design, and several of the most valuable segments of the semiconductor ecosystem. Companies such as Intel defined the microprocessor era, NVIDIA became integral to artificial intelligence and accelerated computing, and AMD emerged as a significant force in processors and high-performance systems. In addition to chip design, the United States held influential roles in semiconductor design software, manufacturing equipment, and research leadership. While the United States did not control every aspect of the supply chain, it retained strong influence over several of the most sophisticated and strategically significant components.

In contrast, China developed its semiconductor industry on a different foundation.

China emerged as a major global hub of electronics manufacturing and, consequently, one of the largest consumers of semiconductors. Its extensive industrial base relies on a continuous supply of chips for smartphones, computers, telecommunications equipment, industrial machinery, appliances, vehicles, and data systems. Even as China became deeply integrated into global electronics production, it continued to rely heavily on semiconductors designed or manufactured outside its borders, particularly for the most advanced chips.

Over time, this reliance came to be viewed by Chinese policymakers as a strategic vulnerability.

In an environment where semiconductors underpin industrial competitiveness and strategic autonomy, heavy reliance on imports constitutes a significant vulnerability.

Chinese policymakers increasingly viewed dependence on foreign semiconductor capabilities as a strategic constraint on advanced-technology leadership. Consequently, China intensified efforts to develop a domestic chip ecosystem capable of competing globally.

This initiative encompassed substantial state-backed investment, industrial policy, research support, and long-term planning. Achieving greater semiconductor self-reliance became a major strategic objective, particularly within broader national initiatives focused on industrial upgrading and technological independence. Programs such as Made in China 2025 identified semiconductors as a critical sector, reflecting their foundational role in economic and geopolitical power.

Chinese firms expanded across various segments of the semiconductor value chain. Semiconductor Manufacturing International Corporation (SMIC) increased its manufacturing ambitions, while other companies invested in design, packaging, materials, equipment, and the development of the domestic ecosystem. China benefited from an extensive market, robust government support, a substantial engineering workforce, and a strong drive for rapid advancement.

Major Chinese technology companies also expanded into chip architecture, with Huawei serving as a prominent driver. Through its semiconductor division, HiSilicon, Huawei developed advanced smartphone processors and other chips. By narrowing down the performance gap with leading mobile chip players in several flagship segments, Huawei reduced its reliance on foreign suppliers for essential hardware components. This development signaled a shift: Chinese firms were moving beyond assembly to the design of increasingly sophisticated components.

This trend attracted increasing attention from policymakers in Washington.

American policymakers increasingly viewed advances in Chinese semiconductor capabilities as strategically significant, particularly because advanced chips can serve both commercial and security-related purposes. These concerns extended beyond smartphones or commercial market competition. Advanced semiconductors are commonly described as dual-use technologies; since the same high-performance computing power that enables artificial intelligence models, cloud services, and industrial automation can also support defense systems, surveillance infrastructure, cyber capabilities, and strategic decision-making tools. In this context, semiconductor capability has become a matter of growing strategic concern.

Consequently, the United States increasingly relied on export controls as a major policy instrument. These regulatory measures were designed to manage the export of sensitive technologies essential to advanced semiconductor production. U.S. policy reflects a strategy of protecting national security interests by implementing rigorous oversight and controls over the most advanced elements of the semiconductor ecosystem.

The export controls targeted several critical vulnerabilities within the semiconductor sector.

One such vulnerability was access to advanced chips. Firms operating in China faced increasing difficulty obtaining high-end processors and accelerators for artificial intelligence, high-performance computing, and other advanced applications. This affected companies that relied on leading-edge semiconductors for products such as smartphones, telecommunications equipment, and data infrastructure.

Another critical vulnerability involved manufacturing equipment. Designing a chip and producing it at the technological frontier are distinct challenges. Modern semiconductor fabrication relies on highly specialized tools, with lithography systems being among the most essential for patterning minute transistor features onto wafers. The most advanced lithography systems are produced by ASML in the Netherlands, whose extreme ultraviolet (EUV) machines rank among the most complex industrial tools ever constructed. Without access to such equipment, manufacturing chips at the most advanced nodes becomes extremely challenging.

This underscores the significance of equipment access. Even with strong engineering talent, robust financial support, and ambitious industrial strategies, limited access to advanced manufacturing tools faces major obstacles in reaching the technological frontier. Semiconductor progress is not linear; it depends on multiple stages of capability distributed across different firms and countries.

Software was also an important factor, particularly electronic design automation (EDA) tools. These specialized platforms are essential for designing, simulating, verifying, and optimizing complex chips containing billions of transistors. Without advanced EDA tools, the design of cutting-edge semiconductors becomes extremely challenging. This underscores that modern semiconductor capabilities depend not only on hardware but also on highly specialized software infrastructure.

As these controls expanded, the semiconductor supply chain increasingly became a focal point of strategic and regulatory competition.

This development represented a significant shift. For years, the global semiconductor industry functioned as a highly interdependent system, with design, manufacturing, equipment production, materials sourcing, and packaging often occurring in different countries. Globalization promoted efficiency, but efficiency and security are distinct considerations. As major powers began treating semiconductors as strategic assets, earlier assumptions about the industry's structure began to erode. The balance between efficiency and resilience has become one of the defining tensions of the modern semiconductor industry.

Countries reassessed their vulnerabilities, and firms reconsidered supply-chain dependencies. Governments questioned the prudence of relying on globally optimized networks that could pose political risks. The semiconductor industry, previously viewed primarily as a commercial sector, began to be discussed in terms of resilience, sovereignty, and strategic control.

The United States responded not only through regulatory measures, but also through industrial policy aimed at strengthening domestic semiconductor capacity. Additional investment was directed toward manufacturing, research, and broader ecosystem development within the United States. While the country remained strong in chip design and several upstream capabilities, a significant share of the most advanced manufacturing capacity had become concentrated overseas, particularly in Taiwan, raising concerns about supply-chain resilience. In response, U.S. policy supported new fabrication facilities, research initiatives, packaging capacity, and related efforts to expand semiconductor production domestically.

China also expanded its domestic semiconductor efforts. Faced with more limited access to certain foreign tools and advanced chips, it increased investment in local foundries, equipment development, software capabilities, materials research, and broader ecosystem building across the semiconductor value chain. Although it remained unclear how far these efforts would narrow existing gaps, the shift toward greater domestic capability and reduced external dependence became increasingly visible.

This dynamic illustrates why the U.S.-China semiconductor relationship has assumed broader strategic significance. It extends beyond tariffs or individual supply contracts and increasingly centers on access to key technologies, manufacturing capacity, software tools, research capabilities, and industrial resilience. In this sense, the semiconductor sector has become a central arena in the wider competition over future technological and economic influence.

The consequences of this conflict ripple across the global semiconductor industry.

Companies operate in an increasingly sophisticated and carefully governed global environment. Partnerships that may once have appeared primarily commercial are also shaped by evolving regulatory frameworks and policy priorities. Supply chains that move across borders require close attention to export controls, licensing requirements, compliance processes, and jurisdiction-specific review mechanisms. As a result, multinational firms often navigate differing legal frameworks, market conditions, and policy expectations across regions. Yet cross-border collaboration remains one of the defining strengths of the semiconductor industry, complemented by a growing emphasis on resilience, regulatory alignment, and long-term strategic planning.

At the same time, demand for advanced semiconductors continues to rise.

Artificial intelligence requires substantial volumes of high-performance computing resources. Data centers are expanding, and electric vehicles are increasing their semiconductor content. Fifth-generation (5G) and future communications systems demand more specialized chips. Edge computing, industrial automation, defense electronics, and advanced consumer devices all require greater capabilities from the semiconductor ecosystem. The global demand is for more advanced and secure chips, even as the political environment surrounding semiconductors becomes increasingly adversarial and fragmented.

For these reasons, semiconductors become crucial, ranking high among the most strategically important technologies globally. They drive economic growth through their foundational role in digital industries, underpin national security by enabling military systems and intelligence capabilities, and determine technological leadership in artificial intelligence, computing, communications, and advanced manufacturing.

The evolving relationship between the United States and China is likely to influence the semiconductor industry for years to come, if not decades. It is expected to influence the location of fabrication facilities, access to critical tools, supply chain design, and national approaches to technological dependence and resilience. It may also shape the broader competitive landscape in areas such as artificial intelligence, advanced computing, and digital infrastructure.

However, the global semiconductor story extends beyond the U.S.-China relationship alone.

Other regions are also seeking to strengthen their positions. Europe is seeking to enhance resilience and expand its semiconductor footprint. Japan is reasserting its role through materials, equipment, and strategic investment. South Korea remains one of the major forces in memory and advanced manufacturing, while India and other emerging technology hubs are striving to establish their own roles within the semiconductor ecosystem. The semiconductor race is not exclusively bilateral, despite the prominence of the US-China relationship.

Nevertheless, the central issue persists.

The significance of semiconductors extends beyond the chips themselves. It increasingly involves access to the technologies, manufacturing capacity, software tools, and industrial capabilities that shape the modern economy. In this environment, questions of dependence, resilience, and technological autonomy have become more visible.

Semiconductors were once regarded simply as components within consumer devices. Today, they are also associated with export controls, industrial policy, security considerations, and expanding regulatory oversight. This evolution highlights the increasingly complex connection between technology, economics, and public policy.

References

1. U.S. BIS 2022, Export Controls on Certain Advanced Computing and Semiconductor Manufacturing Items; Supercomputer and Semiconductor End Use; Entity List Modification.
2. U.S. BIS, 2023, Implementation of Additional Export Controls: Certain Advanced Computing Items; Supercomputer and Semiconductor End Use; Updates and Corrections.
3. U.S. BIS, 2024, Commerce Strengthens Export Controls to Restrict China's Capability to Produce Advanced Semiconductors for Military Applications
4. U.S. BIS, 2025, Commerce Strengthens Restrictions on Advanced Computing Semiconductors to Enhance Foundry Due Diligence and Prevent Diversion to PRC.
5. U.S. BIS 2026, Department of Commerce Revises License Review Policy for Semiconductors Exported to China.
6. U.S. Department of Commerce, Export Controls on Advanced Computing and Semiconductor Manufacturing Items.
7. NIST, CHIPS for America.
8. NIST, A Strategy for the CHIPS for America Fund.
9. NIST, CHIPS R&D Funding Opportunities.
10. CSIS, 2023, China's New Strategy for Waging the Microchip Tech War.
11. CSIS, 2022, The Semiconductor Supply Chain: Assessing National Competitiveness.
12. CSET, *U.S. Semiconductor Exports to China: Current Policies and Trends*.

13. U.S.-China Economic and Security Review Commission, Made in China 2025: Evaluating China's Performance.
14. U.S.-China Economic and Security Review Commission, 2025, Interlocking Innovation Flywheels: China's Manufacturing and Innovation Engine.
15. SEMI, Trade and Export Control.
16. Semiconductor Industry Association, Export Control and National Security Policy.
17. SMIC, 2023, Annual Report.
18. HiSilicon, Kirin 9000 Chipset and Kirin 650 Chipset.
19. ASML, EUV Lithography Systems.
20. Synopsys, What Is Electronic Design Automation.
21. Siemens EDA, Corporate Overview.
22. Intel, Announcing a New Era of Integrated Electronics.
23. NVIDIA, Our History: Innovations Over the Years.
24. AMD Investor Relations, Corporate Overview.
25. The White House, 2022, CHIPS and Science Act.

Chapter 9: Europe's Semiconductor Revival – Rebuilding Capacity in a Fragmented World

For much of the semiconductor age, the global industry appeared to be organized around a clear hierarchy of power. The United States led in chip architecture, design capability, and software ecosystems. Taiwan became highly important through advanced fabrication. South Korea built commanding strength in memory and consumer electronics. Japan retained influence through materials, precision manufacturing, and upstream industrial inputs. In that landscape, Europe was often cast in a quieter role—historically important but less visible in the industry's most prominent competitive arenas.

However, this perception has consistently been more superficial than accurate.

Europe did not disappear from the semiconductor sector; instead, it took on a different role. Rather than dominating all segments of mass production, Europe remained important in several technically demanding and strategically significant layers of the semiconductor ecosystem. Its role became less visible to the public, yet in several respects, it grew more consequential to the system as a whole. In the modern chip industry, visibility does not necessarily equate to power. Some of the most decisive actors are those controlling the tools, materials, and specialist capabilities essential for the industry's functioning, rather than those producing the most prominent symbols of technological prestige.

This distinction is fundamental to understanding Europe's resurgence in the semiconductor industry.

Europe's semiconductor resurgence does not represent a straightforward industrial comeback, nor is it an attempt to replicate the trajectories of Taiwan, South Korea, or the United States. Instead, it reflects a more measured and, in several respects, more illustrative response to the current era. As semiconductors have evolved from commercial components to instruments of national power, the criteria for relevance in this industry have shifted.

Power is increasingly shaped not only by manufacturing volume, but also by control over critical capabilities, resilience during disruptions, and the capacity to secure strategic autonomy in technologies too critical to be left entirely in foreign hands. From this perspective, Europe is not re-entering the semiconductor sector from the periphery; rather, it is reaffirming a significance it never entirely lost.

The most prominent example of this significance is found in the Netherlands.

ASML holds an unusually important position in the modern semiconductor industry. It is the only commercial supplier of extreme ultraviolet lithography systems used for the most advanced chip manufacturing. Lithography is the foundational act of modern chipmaking: the transfer of incredibly precise patterns onto silicon wafers, layer upon layer, until billions of transistors emerge in microscopic form. As each generation of semiconductors pushes toward smaller geometries and higher density, lithography becomes not merely another step in manufacturing, but one of the decisive limits on how far computation itself can advance.

That reality gives ASML a strategic importance that extends far beyond commerce. Its machines are among the most sophisticated industrial systems humanity has ever built, integrating cutting-edge optics, ultra-precise motion systems, powerful light sources, advanced metrology, and engineering tolerances so unforgiving that only a handful of suppliers worldwide can contribute meaningfully to their creation. These are not ordinary capital goods. They are a critical enabling technology at the leading edge of semiconductor manufacturing. They influence who can manufacture the chips that underpin advanced computing, artificial intelligence, defense systems, and future digital infrastructure.

In this context, ASML represents more than a European success story. It exemplifies how a single company, anchored in a specific region, can influence the global balance of technological power. Semiconductor strategies in major economies are influenced by Europe's role in advanced lithography. The most advanced objectives of the digital era depend, at a critical stage, on machinery produced in Europe. This constitutes structural power rather than peripheral relevance.

Nevertheless, Europe's role in the semiconductor sector extends beyond lithography.

The continent also plays a critical role in semiconductor technology, which is becoming increasingly important as the world becomes more electrified, automated, and industrially interconnected. If advanced logic chips are often treated as the brains of the digital era, Europe has remained especially strong in the chips that give machines intelligence and control over physical systems.

STMicroelectronics is an important example of this. Its product base spans microcontrollers, sensors, analog devices, and embedded systems that operate quietly inside the machinery of modern life. These semiconductors do not usually attract headlines, yet they are present in automobiles, industrial equipment, medical devices, smart infrastructure, factory systems, and consumer electronics. They form part of the embedded control layer of the modern economy. As more machines become connected, more environments become sensor-driven, and more industries become software-defined, these kinds of chips move steadily from supporting roles to central positions in the architecture of competitiveness.

Infineon Technologies represents another major pillar of European strength, especially in power semiconductors. These devices sit closer to the movement of energy than to the processing of data. They regulate current, manage conversion, and control the efficient flow of power in electric vehicles, charging systems, industrial motors, renewable energy infrastructure, rail transportation, and grid technologies. They may not command the cultural fascination reserved for AI accelerators or cutting-edge processors, but they are highly important to modern electrified and industrial systems. In a century increasingly shaped by electrification, the energy transition, and climate-driven industrial transformation, power electronics are no longer a peripheral category. They are central to the real economy.

This underscores why Europe's semiconductor position warrants more serious consideration. Europe's strength resides not only in high-profile domains but also in essential, foundational technologies. It is deeply embedded in the technological infrastructure of modern industry.

To say this is not to ignore Europe's losses. In the late twentieth century, as fabrication became more capital-intensive, geographically concentrated, and increasingly supported by aggressive industrial policy in East Asia, Europe's share of large-scale semiconductor manufacturing declined substantially. The center of gravity in semiconductor manufacturing shifted outward, especially toward Taiwan and South Korea, later joined by China's state-backed expansion. Europe remained home to research excellence and specialized industrial leadership, but it no longer served as the primary theater of advanced fabrication growth.

However, decline did not mean disappearance. A crucial part of Europe's manufacturing continuity endured in Dresden, Germany, in what became known as Silicon Saxony. This cluster preserved more than fabrication capacity. It preserved industrial memory, supply relationships, technical confidence, and an ecosystem capable of sustaining advanced semiconductor work inside Europe even as the wider industry globalized around other centers. In the late 1990s, AMD made a strategically important decision to establish major fabrication operations in Dresden. Those investments did not merely create production lines; they helped anchor a semiconductor culture with long-term strategic value.

When AMD later spun off its manufacturing arm to create GlobalFoundries in 2009, Dresden remained one among the most significant fabrication sites in Europe. Over time, Dresden became one of Europe's most significant pure-play foundry locations, linking European research and industrial demand to real manufacturing capability. This continuity mattered far more than it may have seemed at the time. It meant that when semiconductors returned to the center of geopolitical thinking, Europe did not need to build from a condition of total absence. It still had a foothold.

The subsequent direction of GlobalFoundries also reveals an important aspect of Europe's industrial logic. In 2018, the company withdrew from the race to pursue the most advanced process nodes at 7 nanometers and below. On paper, this looked like a retreat. In practice, especially from a European perspective, it marked a strategic shift toward technologies more closely aligned with the continent's economic structure and industrial needs. Specialty nodes such as FD-SOI, along with other process technologies optimized for low power, reliability, and application-specific performance, were particularly well suited to automotive systems, industrial electronics, 5G infrastructure, and the Internet of Things.

This was significant because Europe's semiconductor future was unlikely to depend solely on becoming the world's leading producer of advanced smartphone or data-center logic chips. Its comparative strengths were found in automotive technologies, industrial automation, embedded systems, connectivity, and energy-efficient infrastructure. A foundry model aligned with these sectors did not signify defeat; rather, it acknowledged that industrial relevance does not always require leading the most visible segments. Instead, it involves serving markets that become increasingly important as the physical and digital worlds converge.

The onset of the pandemic provided a stark lesson in the fragility of the global semiconductor system.

For years, the chip supply chain had been praised for its efficiency. Production was distributed internationally, optimized for cost and specialization, and treated as a triumph of global integration. But efficiency is not the same as resilience. When the COVID-19 shock collided with factory disruptions, logistics breakdowns, volatile demand, and inventory imbalances, the system's weakness became unmistakable. In Europe, automotive production lines slowed or halted because of shortages of relatively modest semiconductors, including microcontrollers that would once have seemed too ordinary to command strategic attention.

This period transformed the political significance of semiconductors throughout Europe.

Chips ceased to be viewed as just another imported industrial input. They became visible as critical infrastructure—small, often unseen components whose absence could immobilize factories, delay vehicles, disrupt production chains, and weaken economic sovereignty. For European governments and industries alike, the lesson was profound. A region could possess scientific excellence, engineering depth, advanced companies, and high-value manufacturing, yet remain exposed if too much of its semiconductor dependence rested on fragile external chains.

While the previous model prioritized efficiency, the current era requires resilience.

It is in that context that the European Chips Act should be understood. The initiative is not only an industrial support framework; it is also a resilience and capacity-building strategy. It is a strategic response to a changed geopolitical reality. Europe has recognized that semiconductors now sit at the intersection of economic security, military capability, digital transformation, green transition, and industrial continuity. A technology so deeply woven into every major system of modern life can no longer be treated as a capability that can be outsourced without strategic implications.

The ambition to increase Europe's share of global semiconductor manufacturing reflects this shift in thinking. It is not only about producing more chips. It is about ensuring that Europe retains enough capacity, capability, and strategic control in key areas to reduce excessive dependence on disruptions beyond its influence. In this sense, semiconductor policy has become part of a wider European search for technological sovereignty—a search shaped not by isolationism, but by the recognition that openness without capacity can become vulnerability.

The response has begun to take material form. Major investment plans and industrial partnerships across the continent have signaled an effort to rebuild and deepen Europe's semiconductor base. GlobalFoundries has continued investing in Dresden. STMicroelectronics and GlobalFoundries have advanced their joint expansion in France. Not all of these projects serve the same strategic purpose, nor are they aimed at the same technological tier. Still, together they point toward a broader rebalancing: an attempt to anchor more semiconductor capability within Europe's own industrial domain.

At the heart of this effort lies the idea of sovereign capacity. This does not mean complete self-sufficiency, which is neither practical nor even desirable in an industry as globally interdependent as semiconductors. Rather, it means maintaining sufficient domestic capability in strategically vital areas so that Europe is less exposed to geopolitical pressure, regional instability, or supply disruptions than it would be without domestic capability in key areas. In the semiconductor age, dependence has become more than an economic calculation. It has become a question of strategic exposure.

This context underscores the significance of Europe's semiconductor revival.

It is not a dramatic attempt to dominate every layer of the semiconductor chain, nor a fantasy of replacing Asia as the world's manufacturing center. It is a more realistic and, in many ways, more sophisticated project: to convert Europe's existing strengths in lithography, specialty semiconductors, industrial applications, power electronics, and research excellence into a more secure and influential technological position.

Europe's strength in the semiconductor world remains asymmetrical, but asymmetry is not the same as insignificance. The continent may not match East Asia in sheer fabrication scale, yet it occupies a uniquely important position in advanced lithography through ASML. It holds serious positions in power electronics, embedded intelligence, industrial semiconductors, and automotive-oriented technologies. It benefits from research ecosystems led by institutions such as IMEC, along with dense networks of universities, suppliers, equipment makers, and specialized firms. What is emerging now is an effort to connect those strengths more deliberately to industrial policy, domestic production, and long-term strategic planning.

In this respect, Europe's semiconductor revival is not merely an economic development narrative. It illustrates how power is being redefined in the twenty-first century.

For a long time, Europe's role in semiconductors could be described as quiet competence. That phrase still contains truth, but it no longer captures the full picture. Quiet competence acquires a different meaning when the capabilities involved become globally significant. In a world where technology is increasingly intertwined with diplomacy, defense, trade, and national resilience, Europe's quieter strengths have become more visibly strategic.

The current revival represents a strategic recalibration. Europe is not attempting to emulate other regions in the semiconductor sector. Instead, it seeks to maximize its unique position: a continent that may not dominate every stage of the industry, but whose capabilities are difficult to bypass, difficult to replicate quickly, and increasingly important in the global semiconductor system.

Europe's semiconductor revival is not defined by scale alone, but by the strategic relevance of its capabilities. In a system where influence is increasingly shaped by access to critical technologies, Europe's position illustrates how asymmetry can translate into lasting technological and industrial significance.

References

1. European Commission, 2024, European Chips Act.
2. European Commission, 2022, Digital Sovereignty: Commission Proposes Chips Act.

3. European Parliament, 2022, The EU Chips Act.
4. European Commission, 2023, State Aid: Commission Approves French Measure to Support STMicroelectronics and GlobalFoundries.
5. ASML, The World's Supplier to the Semiconductor Industry.
6. ASML, EUV Lithography Systems.
7. ASML, 2024, 5 Things You Should Know About High NA EUV Lithography.
8. STMicroelectronics, Our Technology Starts With You.
9. STMicroelectronics, Automotive Microcontrollers.
10. STMicroelectronics, 2022, ST and GlobalFoundries to Advance FD-SOI Ecosystem with New 300mm Manufacturing Facility in France.
11. GlobalFoundries, About GF in Europe.
12. GlobalFoundries, 2023, GlobalFoundries and STMicroelectronics Finalize Agreement for New 300mm Semiconductor Manufacturing Facility in France.
13. Infineon Technologies, Semiconductor and System Solutions.
14. Infineon Technologies, Solutions for Electric Vehicles.
15. SEMI Europe, 2025, SEMI Europe – Chips Act Report.
16. SEMI Europe, 2025, European Semiconductor Industry Endorses SEMICON Coalition's Call for a Revised EU Chips Act.
17. WSTS, 2025, Global Semiconductor Market Approaches USD 1 Trillion in 2026.
18. TechInsights, 2023, Chip Observer, October 2023.
19. Silicon Saxony, About Us.
20. imec, About imec.

Chapter 10: India's Semiconductor Ambition – The Quest for Technological Sovereignty

For most of the semiconductor age, India participated in important parts of the semiconductor ecosystem without occupying its most capital-intensive manufacturing tiers. The country became known for software, engineering talent, IT services, embedded systems, and digital entrepreneurship, but not for large-scale semiconductor manufacturing. The world's most advanced fabrication facilities rose elsewhere—in Taiwan, South Korea, the United States, and Japan—while India helped power the digital age from a different position in the value chain. It supplied talent, design support, software integration, and technical capability, yet remained outside the physical core where the most strategic forms of control were concentrated.

While that arrangement was sustainable for a period, it has become increasingly difficult to maintain in the current context.

In the twenty-first century, semiconductors are no longer just industrial components. They are increasingly associated with economic leverage, technological capability, security planning, and geopolitical influence. Chips sit beneath nearly every critical system that defines modern life: telecommunications, banking, cloud infrastructure, automobiles, industrial automation, medical devices, defense electronics, artificial intelligence, surveillance systems, and national digital infrastructure. A country seeking greater technological capability may find that long-term dependence on external semiconductor supply increasingly constrains the physical foundations of its digital economy.

This context forms the backdrop to India's semiconductor ambition.

India's interest in semiconductors is not simply about industrial modernization, nor is it merely about participating in a fashionable global policy trend. It is about recognizing a harder strategic truth: in a world increasingly shaped by digital systems, sovereignty is no longer defined only by territory, military force, or monetary power. It is also shaped by access to computing, control over supply chains, domestic technological capability, and the capacity to participate in the industries that underpin all other advanced industries. Countries that remain dependent on others for semiconductors risk remaining dependent at the very moment when dependence has become more strategically consequential.

The global semiconductor system has revealed this reality with unusual clarity. For years, the industry was celebrated as one of the most sophisticated examples of globalization. Design could happen in one country, software in another, lithography equipment in Europe, specialty materials in Japan, wafer fabrication in Taiwan, memory in South Korea, packaging in Southeast Asia, and final assembly somewhere else again. This distributed architecture achieved extraordinary efficiency but also created concentration, fragility, and strategic vulnerability. When supply disruptions struck in the early 2020s, triggered by the COVID-19 pandemic, the illusion of seamless resilience began to collapse. Automakers slowed production. Consumer electronics experienced delays. Governments rediscovered that modern industrial power cannot rest entirely on distant, tightly concentrated supply chains.

India's policy response increasingly reflected these concerns.

For a country of India's size, population, market scale, digital ambition, and geopolitical aspirations, semiconductor dependence is not a minor commercial inconvenience. It can create a strategic vulnerability. A country that seeks leadership in electronics manufacturing, digital public infrastructure, defense modernization, electric mobility, artificial intelligence, telecommunications, and critical infrastructure cannot permanently outsource the physical foundations of those ambitions. Software strength matters. Market size matters. Talent matters. But sooner or later, serious national ambition runs into silicon.

Consequently, semiconductors have shifted from the periphery of Indian industrial policy to the center of strategic planning.

India's semiconductor ambition must therefore be understood not as a narrow manufacturing program, but as part of a larger reappraisal of national power. What is emerging is less an industrial programme and more a redefinition of how technological capability is understood at a national level. A country that cannot secure access to semiconductors—or build meaningful capacity within the ecosystem—risks finding that many of its other ambitions rest on fragile external dependence.

The global semiconductor power game has intensified this shift. Major economies, including the United States, China,

Europe, Japan, and South Korea, have all intensified semiconductor-related industrial and strategic efforts. In such an environment, semiconductors are no longer just commercial products moving through markets. They are increasingly part of the architecture of power.

India is increasingly affected by that contest, even when it is not a primary participant in every aspect of it.

Even countries that are not direct participants in the fiercest technological rivalries are shaped by them. If the future digital economy is being structured by those who control chip design, fabrication, packaging, materials, tools, and supply chains, then countries that remain outside those layers risk having less influence over the technologies that will define the century. India's semiconductor ambition is, in part, an attempt to avoid strategic a structurally peripheral position. It reflects the recognition that a country of India's scale may seek to reduce long-term dependence as only a downstream customer in one of the most consequential industries of the age.

India's entry into the semiconductor market is distinctive, as the country does not begin from a position of inexperience. Its foundation is rooted in a longstanding tradition of scientific institution-building, strategic electronics, and unrealized industrial potential. Prior to the centrality of semiconductors in industrial policy, India had already developed expertise in physics, materials science, instrumentation, and mission-critical electronics through institutions such as CSIR, CEERI, BARC, DRDO, and ISRO-affiliated ecosystems. Although this early foundation did not yield a globally competitive chip industry, it established significant technical depth, institutional memory, and a national orientation toward technological self-reliance.

India's trajectory in the semiconductor sector was not historically absent but rather incomplete. The Semiconductor Laboratory (SCL) at Mohali provided an early entry point into semiconductor manufacturing. SCL's origins predated TSMC's founding in 1987, although that early beginning did not result in comparable industrial scale or ecosystem development. In its formative phase, SCL collaborated with American Microsystems Inc. (AMI) and advanced its CMOS process capabilities from a 5-micron node and progressed through smaller technology nodes through sustained research and development.

However, chronological precedence did not translate into industrial strength. India's initial progress was undermined by disruptions, notably the 1989 fire at the SCL facility, and more fundamentally by the lack of ecosystem discipline, sustained capital, supplier networks, and manufacturing scale that enabled East Asian countries to achieve global leadership. India's early semiconductor efforts illustrate that capability without continuity rarely translates into industrial leadership.

Consequently, India entered the semiconductor era, occupying a different segment of the value chain. From the 1990s onward, the country became prominent in design, embedded systems, verification, validation, engineering services, and software-driven chip development, while advanced fabrication remained concentrated elsewhere. Multinational semiconductor firms established major centers in Bengaluru, Hyderabad, Noida, Pune, and other cities. Engineers in India contributed to chip design workflows and system development, even if the most advanced manufacturing happened elsewhere. This distinction is significant.

This resulted in a significant imbalance: India contributed intellectual value to the semiconductor industry but did not participate in its manufacturing core. In a world where semiconductors shape economic resilience, military capability, technological sovereignty, and geopolitical leverage, continued dependence may become increasingly difficult to reconcile with India's long-term strategic objectives. India's semiconductor ambition is therefore best understood not as a narrow manufacturing project, but as an attempt to close a historical gap—between talent and fabrication, participation and control, capability and power.

Semiconductor power does not come only from fabs. It also comes from human capital, technical depth, design capability, research institutions, and ecosystem linkages. India's established design base gives it a stronger point of departure than countries entering the industry with little prior involvement. It already has engineers, multinational relationships, technical institutions, and a role in the global semiconductor ecosystem. But design strength and manufacturing strength are not the same thing. A country can help develop chips used around the world and remain dependent on foreign manufacturing for their physical realization.

At its core, India's semiconductor ambition seeks to close this strategic gap.

That gap is not merely industrial. It is geopolitical. Countries with strong semiconductor manufacturing positions often enjoy advantages in resilience, bargaining power, and strategic partnerships. They influence access, shape supply chains, attract talent, build strategic partnerships, and acquire a kind of quiet leverage that is difficult to replicate elsewhere. Semiconductor capability strengthens national resilience by making a country less exposed to shocks beyond its control. It also strengthens bargaining power because participation in a critical industry change how others engage with it.

This is one reason the semiconductor industry has become a site of industrial policy rather than pure market logic. Many major economies no longer treat semiconductors as ordinary traded goods. They are too central to economic security, technological competitiveness, and national defense. Once that reality becomes clear, the logic of state involvement changes. Incentives, subsidies, long-term coordination, public-private partnerships, ecosystem planning, and institutional support are no longer exceptional. They increasingly appear justified in the eyes of policymakers.

India's semiconductor program should be interpreted within this broader context.

The India Semiconductor Mission is the principal institutional vehicle for this strategic shift. It reflects the recognition that semiconductors are not an industry built solely on rhetoric, nor one left entirely to market forces. Semiconductor manufacturing is among the most demanding industrial activities on Earth. It requires enormous capital, reliable infrastructure, specialized materials, clean-room environments, technical labor, supplier ecosystems, precision engineering, process discipline, and years of accumulated learning. Even wealthy countries do not build serious semiconductor capabilities casually.

India is not pursuing semiconductors because entry is easy. It is pursuing them because staying outside has become strategically harder to justify.

However, ambition and capability are distinct concepts.

India faces real constraints. Semiconductor manufacturing cannot be created simply through financial incentives or policy declarations. It requires stable electricity, clean water, chemical-handling systems, logistics, institutional continuity, and deep pools of specialized skills. It requires industrial partners willing to commit to long timelines, absorb complexity, and navigate the uncertainty of returns. It requires the central and state governments to coordinate, which is often difficult in practice. It requires not just attracting projects, but sustaining them. And above all, it requires realism. No country can enter the semiconductor industry at scale and instantly become Taiwan, South Korea, or the United States. The industry rewards sustained execution and tends to expose weakly supported ambitions over time.

For this reason, India's semiconductor ambition relies a serious and pragmatic approach rather than an idealized perspective.

Its significance lies not in the fantasy of immediate transformation, but in the fact that India has now recognized semiconductors as a field in which it must compete, not merely participate from the sidelines. That recognition is itself important. It marks a change in how India understands technological power. Chips are no longer being viewed only as imported inputs for electronics manufacturing or as distant products made elsewhere. They are being understood as strategic assets that shape economic resilience, technological autonomy, industrial depth, and geopolitical relevance.

This shift in mindset may prove significant alongside any individual investment.

It suggests that India is beginning to think of semiconductors not just as an industry, but as a layer of national capability. Such a shift affects how the country approaches education, industrial policy, foreign partnerships, infrastructure, and technological priorities. It also affects how India positions itself vis-à-vis other powers. A country that expands its semiconductor role may also expand its influence within the evolving technology economy.

And yet India's ambition is not solely defensive. It is not only about reducing vulnerability. It is also about claiming opportunity.

Semiconductors offer India a chance to align several of its long-term objectives: manufacturing growth, deeper technological capability, higher-value industrialization, strategic relevance, electronics expansion, skilled employment, and greater participation in critical supply chains. In that sense, semiconductor ambition is not disconnected from the broader Indian developmental project. It is one of the more visible areas where economic policy, industrial strategy, and national security considerations intersect.

This perspective underscores that the issue extends beyond semiconductors alone.

India's semiconductor ambition is ultimately to be more than a downstream user of advanced technology. It is an ambition to gain a more meaningful place in the systems that create technological power in the first place. Whether India succeeds fully, partially, or unevenly remains to be seen. But the strategic logic behind the effort is already clear. A country of India's scale is increasingly unlikely to remain absent indefinitely from one of the industries that shape the global distribution of power.

In this sense, India's semiconductor ambition is not just about fabs, packaging plants, or policy incentives. It is about the recognition that the future balance of power will depend not only on who writes code, owns data, or builds digital platforms, but also on who can secure a place in the physical technological foundations of the modern world.

India's recent policy choices indicate that it seeks such a place.

Whether it can secure one will depend on execution, patience, ecosystem-building, and time. But in the global semiconductor power game, the first decisive move is often not mastery. It is the decision to enter the contest with seriousness.

The recent semiconductor policy direction at both national and state levels suggests that a decision has now been made.

References

1. India Semiconductor Mission, About ISM.
2. India Semiconductor Mission, Vision and Objectives.
3. India Semiconductor Mission, Semicon India Programme.
4. India Semiconductor Mission, Compound Semiconductor Scheme.
5. India Semiconductor Mission, SEMICON India 2025.

6. Press Information Bureau, 2024, Semicon India Programme.
7. Press Information Bureau, 2026, India Semiconductor Mission 2.0.
8. Press Information Bureau, 2026, Semicon India Programme Advances with Approval of 10 Semiconductor Units.
9. Press Information Bureau, 2025, India's Semiconductor Revolution.
10. Press Information Bureau, 2025, Powering the Future: The Semiconductor and AI Revolution.
11. Press Information Bureau, 2024, A Game-Changer for India's Semiconductor Ecosystem.
12. HCIKL, 2024, New India Opportunities for Semiconductor Manufacturing.
13. Kapoor, D., 2007, Building National Digital Infrastructure: A Case Study of Semiconductor Complex Limited, Business Review.
14. NITI Aayog, 2024, Powering India's Participation in Global Value Chains.
15. NITI Aayog, 2025, Reimagining Manufacturing: India's Roadmap to Global Leadership in Advanced Manufacturing.
16. NITI Aayog, 2025, Beyond Planning: India's Urgent Need for a 10-Year R&D Vision, Action, and Accountability.
17. Information Technology and Innovation Foundation, 2024, India Semiconductor Readiness.

PART III: The Technology Frontiers

Chapter 11: The Limits of Moore's Law – When Physics Pushes Back

For over fifty years, the semiconductor industry followed a remarkably consistent trajectory. Each successive generation of chips became smaller, faster, denser, and more powerful than its predecessor. Computing advanced through a near-continuous progression rather than sporadic leaps. Engineers consistently increased the capability while reducing the cost and size of each transistor. This period represents one of the most sustained episodes of technological advancement in modern history.

This consistent progression became known as Moore's Law.

First articulated in 1965, Moore's Law described the steady doubling of transistor density alongside declining cost per function. Over time, it evolved from an observation into the organizing principle of the semiconductor industry, enabling sustained exponential improvements in computing capability and shaping the modern digital economy.

The impact of this progress was transformative. Computing power increased exponentially, enabling devices to shrink from room-sized machines to desktops, laptops, phones, wearables, and ultimately to embedded systems integrated into everyday infrastructure. Entire industries emerged as a result. Personal computing became feasible due to more powerful and affordable processors. The ubiquity of smartphones was driven by dense, efficient chips that provided significant capability in compact devices. Cloud computing expanded as servers delivered greater performance per rack, per watt, and per dollar. Moore's Law was not simply a trend in chip manufacturing; it was a fundamental force shaping modern civilization.

However, exponential progress inevitably encounters fundamental limitations.

By the early twenty-first century, semiconductor engineers were encountering increasingly severe physical and economic constraints on traditional scaling. Transistors had become so small that foundational assumptions from earlier generations no longer held. At these dimensions, engineering challenges extended beyond classical electrical behavior to include quantum effects, heat density, material constraints, and economic pressures. Sustaining further miniaturization became increasingly difficult.

One immediate challenge was leakage current. Ideally, a transistor functions as a precise switch, activating only when required. However, at nanometer scales, this behavior becomes difficult to maintain. Electrons may tunnel through barriers that previously blocked them, resulting in unwanted current even when the transistor should be off. This phenomenon wastes energy and reduces efficiency. In modern chips with billions of transistors, even minor leakage effects can have substantial consequences.

Thermal management presents another significant challenge.

Each time a transistor switches, it generates heat as a byproduct. In chips containing billions of transistors operating at high frequencies, managing this heat becomes a significant engineering challenge. Increased performance typically results in greater heat output, which complicates cooling, imposes stricter design constraints, and necessitates trade-offs among speed, power, and reliability. This thermal limitation contributed to the stagnation of clock speed increases, as physical constraints became more pronounced.

The economics of semiconductor advancement also became more challenging. Previously, scaling provided multiple benefits: increased transistor counts, improved performance, and often reduced cost per function. However, as fabrication moved toward atomic-scale precision in critical features, the cost of maintaining technological leadership increased dramatically. Modern fabrication facilities require investments of tens of billions of dollars, and the equipment involved is among the most complex ever constructed. Process development has become slower, riskier, and more expensive. Thus, even when further transistor shrinkage is technically feasible, it is no longer straightforward from an economic standpoint. The deceleration of Moore's Law is attributable to both physical and financial constraints, rather than a lack of engineering effort.

Advanced lithography has been a central technology that has sustained industry progress. The capability to print increasingly smaller features onto silicon wafers has been essential for each major scaling advance. The semiconductor industry demonstrated significant ingenuity in this area. ASML's EUV lithography systems extended scaling beyond the limits of previous lithography approaches. These systems integrate advanced light sources, optics, precision stages, and vacuum technologies, representing a remarkable engineering achievement.

However, even EUV lithography does not resolve all underlying challenges. While it extends the technological frontier, further transistor miniaturization remains increasingly expensive, complex, and constrained by physical limitations. Consequently, many observers have argued that Moore's Law is slowing or approaching its end.

This assessment is partially accurate but does not capture the full picture.

If Moore's Law is interpreted solely as continued transistor shrinkage, then the traditional model is indeed under pressure. However, the semiconductor industry's resilience lies in its capacity to adapt when a particular approach becomes unsustainable. As straightforward scaling became more difficult, engineers responded by redefining the challenges they sought to address.

A significant development involved redesigning the transistor itself.

Traditional transistors were planar, constructed as flat structures on the silicon surface. As controlling current became more challenging at smaller geometries, the industry transitioned to three-dimensional transistor architectures. Fin Field-Effect Transistors (FinFETs) exemplify this shift. Rather than a flat channel, the transistor extends above the surface, enabling the gate to wrap around it and provide enhanced electrical control. This innovation reduced leakage and maintained performance as device dimensions continued to decrease.

FinFETs extended the viability of continued scaling, which is critical in the semiconductor industry. Each new generation of scaling enables the development of new devices, platforms, and markets. More recently, the industry has advanced toward designs such as gate-all-around transistors, where the control structure completely surrounds the channel. These architectures demonstrate that semiconductor progress increasingly relies on device innovation rather than solely on miniaturization.

Leading manufacturers, including TSMC, Intel, and Samsung, have made substantial investments in these technological transitions. Their efforts illustrate a broader industry principle: when one approach reaches its limits, further progress is achieved through structural innovation. The continued evolution of the transistor exemplifies this adaptive strategy.

Another significant strategy has involved redefining the concept of a chip.

Historically, the semiconductor industry aimed to integrate as much functionality as possible onto a single monolithic die. However, as manufacturing complexity increased and large dies became more difficult and costly to produce efficiently, engineers adopted modular approaches. Functions could be divided into smaller components, known as chiplets, which are then integrated into a single package.

Although this approach may appear to be a compromise, it often provides significant advantages.

Chiplets enable different system components to be manufactured separately, potentially utilizing distinct process nodes or optimized technologies, before integration into a single high-performance package. This approach enhances flexibility and can reduce both cost and manufacturing risk. It also allows companies to combine functions more efficiently, rather than consolidating all features onto a single large die. AMD has made chiplet-based design a major part of its processor strategy, using chiplets to achieve high performance while managing design and fabrication complexity.

This shift is significant because it alters the primary drivers of progress. Rather than relying exclusively on transistor density, performance improvements now derive from enhanced integration, packaging, interconnects, and advanced architectural decisions. The industry is transitioning from a focus on miniaturization to an emphasis on systems engineering.

The exploration of new materials represents another dimension of this broader adaptation.

Silicon has served as the foundation of modern semiconductor electronics for decades due to its abundance, well-understood properties, and robust industrial support. However, silicon is not optimal for all applications. Researchers and companies are increasingly investigating alternative materials such as gallium nitride and silicon carbide, particularly for power electronics, electric vehicles, charging systems, and industrial energy management. These materials can offer performance advantages under high-voltage, high-temperature, or high-frequency conditions.

Graphene and other advanced materials have attracted interest for their promising electrical properties, although their commercial applications are still developing. The key consideration is not the imminent replacement of silicon, but rather the emergence of a more diverse material landscape in which various applications leverage distinct physical advantages.

Even more transformative possibilities are being explored.

Quantum computing may eventually address problem classes that are not well-suited to classical transistor-based systems. Photonic processors, which utilize light instead of electrons for specific operations, could provide speed or efficiency benefits for specialized workloads. Neuromorphic computing seeks to emulate aspects of biological neural structures. While these approaches are not yet ready to supplant mainstream digital chips, they indicate that the future of computing may involve multiple architectures.

Therefore, it is essential to accurately interpret the significance of Moore's Law.

Fundamentally, Moore's Law was not solely about increasing transistor counts. It represented sustained technological advancement driven by engineering innovation. The principle embodied the notion that the computing industry could continually achieve greater efficiency by overcoming limitations through creative redesign. This innovative spirit persists, even as the specific mechanisms evolve.

The deceleration of traditional scaling may be prompting the industry to enter a more innovative phase. Previously, progress was primarily achieved by reducing transistor size. Currently, advancements increasingly result from improvements in architecture, packaging, integration, materials, and software-hardware co-design. Although these approaches are more complex, they may yield greater long-term benefits. As one dominant strategy wanes, multiple new avenues for progress often emerge.

Artificial intelligence is a major force accelerating this transition.

AI workloads have revealed both the strengths and limitations of conventional scaling. These applications require substantial computational throughput, high memory bandwidth, efficient data movement, and increasingly specialized architectures. As a result, the industry has shifted toward GPUs, AI accelerators, advanced packaging, high-bandwidth memory, and system-level design strategies that extend beyond the traditional Moore's Law framework. AI is not merely a beneficiary of semiconductor innovation; it is actively reshaping industry priorities.

This shift may become a defining development of the next era in semiconductors.

The era of straightforward transistor miniaturization is fading, while semiconductor innovation continues through other routes.

The industry is increasingly required to adopt inventive, architectural, and system-oriented approaches. Future breakthroughs are likely to result not from further miniaturization of individual transistors, but from reimagining how billions of switches, multiple dies, novel materials, and specialized workloads are integrated into new computing paradigms.

The limits of Moore's Law do not mark the end of semiconductor progress, but a transition in how progress is achieved. As physical scaling slows, innovation increasingly moves toward architecture, integration, and system design. In this new phase, the industry is not reaching an endpoint—it is redefining its trajectory.

References

1. Intel, Moore's Law.
2. Moore, G. E., 1965, Cramming More Components onto Integrated Circuits.
3. Nature Nanotechnology, 2023, Moore Must Go On.
4. Nature Electronics, 2023, The Next Generation of Gate-All-Around Transistors.
5. Nature, 2014, Limits on Fundamental Limits to Computation.
6. Intel Foundry, Semiconductor Manufacturing Process | Intel 18A, 3, and 16.
7. TSMC, 7nm Technology
8. TSMC, 2nm Technology.
9. ASML, EUV Lithography Systems.
10. AMD, 2024, Chiplet Ecosystem White Paper
11. AMD, CDNA 4 Architecture White Paper.
12. AMD Threadripper Pro White Paper.
13. AMD, Zen Core Architecture.
14. NVIDIA, A100 Tensor Core GPU Architecture.
15. Samsung Semiconductor, GAA Structure Transistors.
16. Nature Electronics, 2024, Advanced Packaging of Chiplets for Future Computing Needs.
17. Nature Electronics, 2024, High-Performance, Power-Efficient Three-Dimensional System-in-Package Designs with Universal Chiplet Interconnect Express.
18. Nature Reviews Materials, 2022, 2D Semiconductors for Specific Electronic Applications.
19. IBM Research, Photonic Computing.
20. IBM, What Is Quantum Computing?

21. NVIDIA, Data Centers for the Era of AI Reasoning.

Chapter 12: The Rise of AI Chips – Forging the Engines of Machine Intelligence

In recent years, artificial intelligence has moved from the margins of computer science to the center of the technology industry. It now shapes everything from voice assistants and recommendation engines to medical imaging, autonomous systems, fraud detection, language models, and scientific research. AI is no longer a niche tool used by specialists in laboratories. It has become one of the most influential forces reorganizing the digital economy.

At the core of this transformation is silicon.

The rise of artificial intelligence has not happened solely in software. However dazzling the algorithms may appear, they depend on hardware capable of carrying out an enormous volume of computation. Machine learning, especially modern deep learning, is hungry in a way that few previous computing workloads were. It demands speed, parallelism, memory bandwidth, and power efficiency at a scale that has forced the semiconductor industry to rethink what a processor should look like. The result has been the emergence of a new class of chips designed not just to run computers in general, but to run AI workloads at an industrial scale.

To understand why this shift matters, it helps to start with the traditional processor. For decades, the central processing unit, or CPU, was the undisputed workhorse of computing. CPUs remain highly effective for general-purpose tasks. They handle sequential instructions, switch efficiently between different workloads, and manage the overall logic of a computer system. They are versatile, disciplined, and essential. But versatility comes with trade-offs. Many AI workloads are not primarily about executing one complex instruction after another. They involve enormous numbers of similar mathematical operations repeated across massive datasets. Neural network training, for example, requires the system to repeatedly perform countless matrix multiplications, tensor operations, and parallel calculations.

Such computational demands require a fundamentally different architecture.

Instead of a processor optimized mainly for sequential control, AI workloads thrive on parallelism. They benefit from hardware that can perform thousands of calculations simultaneously rather than fewer, more flexible operations performed one after another. This is where the modern wave of specialized AI-chip development began to take clearer shape.

One of the most commercially significant turning points came from NVIDIA. Originally known as a graphics company, NVIDIA developed graphics processing units (GPUs) for video games and other visual computing applications. These chips were designed to handle large numbers of calculations in parallel because rendering graphics requires vast numbers of similar operations to be performed at once. It turned out that this same characteristic made GPUs extraordinarily well-suited to machine learning.

This development represents a rare instance in technological history where a tool designed for one application became transformative in another. GPUs, originally intended for rendering complex graphics, evolved into one of the principal hardware engines for modern artificial intelligence training. Their thousands of smaller processing cores enabled significantly more effective neural network training compared to conventional CPUs. As AI research accelerated during the 2010s, GPUs rapidly became the leading platform for machine learning training, shifting from an alternative architecture to the default infrastructure of the AI boom.

Soon, the world's largest technology companies began filling data centers with GPU-powered servers. These machines could train increasingly complex models on enormous datasets, pushing AI capabilities further each year. Image recognition improved. Natural language processing surged ahead. Recommendation systems became more sophisticated. Generative AI emerged as a serious force. Beneath all of it sat rows of high-performance chips, drawing enormous amounts of power and performing staggering volumes of computation. The software captured headlines. The silicon made it possible.

As demand for AI hardware grew, competition intensified across the semiconductor industry.

AMD and Intel, long major names in processors and computing infrastructure, moved more aggressively into AI accelerators and high-performance computing. Each recognized that the future of computing would not be defined only by general-purpose CPUs, but by specialized systems built for machine learning and data-intensive workloads. The AI hardware market was too strategically significant for major incumbents to ignore.

At the same time, the rise of AI pushed major technology firms to become chip designers themselves. Google was one of the earliest large-scale examples. Rather than relying entirely on off-the-shelf processors, it introduced the Tensor Processing Unit (TPU), a custom chip designed specifically to accelerate machine learning tasks within Google's own infrastructure. This marked a deeper shift in the industry. It was no longer enough to buy generic computing hardware and hope it would suit every need. The largest technology firms began realizing that if AI was central to their future, then the hardware running AI had to be tuned precisely to their own models, services, and scale.

Other large cloud and platform companies increasingly adopted a similar approach. Cloud providers, hyperscalers, and platform giants increasingly pursue custom silicon for training, inference, networking, and data-center efficiency. This reflected a broader redesign of computing priorities. General-purpose processors still mattered, but the era of one-size-fits-all hardware was fading. In its place came a new philosophy: build the chip around the workload.

This trend is commonly referred to as domain-specific computing.

The idea is straightforward and powerful. Instead of designing hardware that tries to handle every possible task reasonably well, engineers create processors optimized for specific categories of computation. In AI, that might mean chips tuned for neural network training, inference, image recognition, transformer models, or edge-based intelligence. By specializing in hardware, designers can often achieve dramatic gains in performance, efficiency, and cost-effectiveness. The chip becomes less universal but far better at the task that matters most.

That change has opened the door to a wave of innovation, including from startups. A number of younger semiconductor companies have emerged with architectures aimed specifically at machine learning workloads. That distinction is important. Training is the process of building and refining a model, often done in massive data centers using huge volumes of compute. Inference is the act of applying that trained model to real-world data. It happens when a phone interprets speech, a camera identifies an object, a sensor detects an anomaly, or a vehicle reacts to its surroundings. Inference often needs to be fast, efficient, and performed in real time, sometimes on devices with tight energy limits.

This shift creates opportunities for the development of diverse chip architectures.

An inference chip in a smartphone has very different constraints than a training accelerator in a hyperscale AI cluster. One may need to be small, cool, and miserly with power. The other may be built for raw throughput and operate in a power-hungry data center environment. This diversity of use cases has encouraged architectural experimentation across the industry. AI hardware is no longer one market. There are many overlapping markets with different demands, from cloud-scale training to industrial edge deployment.

This development also contributes to the emergence of another significant trend: the rise of edge AI.

For years, much of the AI industry focused on centralized data centers. Models were trained in large facilities and often deployed through the cloud. But many AI applications benefit from running directly on local devices. Smartphones increasingly process voice, vision, and language tasks on-device. Drones need rapid decision-making without waiting for distant servers. Robots and autonomous systems must respond in real time. Smart cameras, industrial sensors, and connected machines often need local intelligence for speed, privacy, and reliability.

That shift toward edge computing has major implications for the semiconductor industry.

Running AI at the edge requires chips that are not only capable but also compact and energy-efficient. They must deliver meaningful performance in devices with limited space, limited cooling, and often limited battery life. This has pushed semiconductor makers to develop embedded AI processors tailored for edge environments. In these systems, latency matters enormously. In applications such as autonomous driving and industrial monitoring, local response time can be critical. A smart factory sensor cannot afford to delay if it is detecting a dangerous anomaly. Edge AI chips bring intelligence closer to the action.

These advancements also illustrate a broader reality in computing: intelligence is becoming increasingly distributed.

It is no longer confined to giant data centers. It is moving into everyday devices, infrastructure systems, industrial environments, and physical machines. That makes AI hardware not just a cloud story, but a system-wide semiconductor story.

Meanwhile, at the high end of the AI spectrum, computational demand continues to surge. Training state-of-the-art AI models often requires substantial hardware resources. Some models involve hundreds of billions, or even trillions, of parameters, and training them can often require thousands of advanced processors working together across massive data center clusters. These systems are not just collections of chips. They are tightly orchestrated computational platforms that include accelerators, high-bandwidth memory, networking fabrics, cooling systems, and software stacks, designed to squeeze out every possible unit of performance.

Consequently, artificial intelligence is driving semiconductor technology to its operational limits.

It is not enough to make chips faster in isolation. AI training systems demand performance across the entire architecture. Memory bandwidth must be higher. Interconnects between processors must be faster and more efficient. Packaging must become more advanced so that logic and memory can sit closer together. Power delivery must be improved. Thermal management becomes critical. In many AI systems, the challenge is no longer just the strength of one processor. It is the efficiency of the whole computational organism.

That has made advanced packaging and system integration increasingly important. High-bandwidth memory, chip-to-chip interconnects, multi-die systems, and increasingly sophisticated substrate and packaging technologies have become central to AI hardware design. Foundries such as TSMC play a critical role here, not only by fabricating advanced processors but also by enabling the manufacturing and packaging methods needed to turn them into deployable AI systems. The modern AI boom can be considered as much a triumph of semiconductor engineering and industrial coordination as it is of algorithmic discovery.

The rise of AI chips is also changing how the semiconductor industry thinks about value.

For years, chip progress was often framed in terms of general-purpose computing. Faster CPUs, denser memory, smaller process nodes. Those things still matter. But AI has shifted attention toward workload-specific advantage. In many AI markets, the most commercially valuable chip is not necessarily the most general-purpose one. It is the one that delivers the best performance for a given class of computation, whether that means training a giant model in a cloud cluster or performing efficient inference inside a mobile device.

That has altered competitive dynamics across the industry. It has elevated the importance of companies with strong AI hardware platforms. It has encouraged cloud providers to become chip designers. It has created room for startups with novel architectures. It has forced legacy semiconductor firms to adapt. In short, AI has not just created a new product category. It has materially reshaped competitive priorities across the semiconductor industry."

The relationship between artificial intelligence and semiconductor development is bidirectional.

Artificial intelligence is not only a driver of semiconductor demand. It is beginning to influence semiconductor design itself. Engineers are using AI tools to optimize chip layouts, model performance behavior, improve verification, and accelerate parts of the design process. This creates a fascinating loop: advanced chips make modern AI possible, and AI now helps design more advanced chips. One generation of silicon helps shape the next. Humanity does enjoy building recursive systems and then pretending this is perfectly normal.

As computing demands continue to rise, the industry is being pushed toward new architectural approaches. Monolithic scaling alone is no longer enough. System-level design matters more. Specialized accelerators matter more. Advanced packaging matters more. Memory and interconnect strategy matter more. AI has become one of the strongest forces driving these changes.

That is why the rise of AI chips is about more than one successful company or one fashionable market category. It signals a deeper transition in computing itself. The age of purely general-purpose dominance is giving way, at least in part, to an age of heterogeneous architectures built around specific workloads. Intelligence is becoming a design target. The chip is no longer just the engine beneath software. It is being shaped directly by the demands of machine learning, data movement, and large-scale model execution.

This transformation remains in its early stages.

As AI systems become more powerful, more embedded, and more widely distributed, the need for new semiconductor architectures will only grow. Some of the next advances will come from faster processors. Some will come from better memory integration. Some will come from domain-specific design. Some will come from new ways of linking multiple chips into larger systems.

One of the most promising of those paths is the rise of chiplets, modular semiconductor components that can be combined into powerful integrated systems.

A single, monolithic chip is unlikely to define the entire future of AI hardware.

Instead, it may be characterized by multiple smaller components operating in concert with high precision.

The rise of AI chips reflects a broader transformation in computing. As intelligence becomes a primary design target, semiconductor innovation is increasingly shaped by the demands of machine learning, data movement, and system-level coordination. In this evolving landscape, silicon is no longer a supporting layer—it is an active driver of how intelligence is created and deployed.

References

1. NVIDIA, 2026, CUDA Platform for Accelerated Computing.
2. Google Cloud, 2026, Cloud Tensor Processing Units.
3. Google Cloud, 2026, TPU Architecture.
4. Google Cloud, 2026, TPU v4.
5. Nature Electronics, 2023, AI Hardware Has an Energy Problem.
6. TSMC, 2026, CoWoS.
7. TSMC, 2026, Advanced Packaging Services.
8. TSMC, 2024, Advancing 3D IC Design for AI Innovation.
9. Samsung Semiconductor, 2026, HBM3E.
10. AMD, 2026, AMD Instinct MI300 Series Accelerators.
11. Intel, 2024, Intel Gaudi 3 AI Accelerator White Paper.
12. Qualcomm, 2026, Qualcomm AI Products.

13. Apple Machine Learning Research, On Device Llama 3.1 with Core ML.
14. Arm, 2026, Ethos-U85.
15. UCIe Consortium, 2026, Specifications.

Chapter 13: Advanced Lithography – Machines That Print the Future

As semiconductor devices approach atomic-scale dimensions, a primary challenge emerges: patterning features that are sufficiently small, precise, and densely packed to enable continued advancement in chip performance. While the semiconductor industry frequently highlights design breakthroughs, transistor architectures, and packaging innovations, these achievements are only meaningful if the corresponding structures can be accurately printed onto silicon. Ultimately, advanced computing relies on the precise transfer of abstract circuit designs onto wafers with exceptional accuracy.

This task is accomplished through lithography.

Lithography is among the most critical processes in semiconductor fabrication, as it determines the minimum feature size, packing density, and the extent to which performance, efficiency, and density can be advanced. In contemporary semiconductor manufacturing, ASML, a Dutch company, has become exceptionally important due to its leading lithography systems, which are essential for producing advanced chips.

ASML occupies a unique and strategically significant position within the global technology supply chain. Although the company does not produce consumer devices or receive widespread public attention, its lithography machines are essential for manufacturing advanced processors used in artificial intelligence, smartphones, high-performance computing, and autonomous systems. ASML's role as one of the most critical suppliers is underscored by its uniquely important position over a key technological chokepoint.

Understanding this significance requires an examination of lithography's fundamental principles.

Lithography is fundamentally a sophisticated printing process. A chip design is encoded onto a patterned mask, or reticle, and projected onto a silicon wafer coated with a light-sensitive material called photoresist. The light-transferred image defines the microscopic features that will become transistors, interconnects, and other circuit structures. While the process may appear straightforward in description, it represents one of the most demanding feats of industrial precision.

The lithography process consists of several meticulously controlled stages. Initially, the wafer is coated with a thin layer of photoresist, a chemical that reacts to light exposure. The mask containing the circuit pattern is aligned within the lithography system, and light is projected through or reflected from the pattern onto the wafer surface. The exposed photoresist is then chemically developed, leaving the desired image. This pattern is subsequently transferred into the underlying materials through etching and additional process steps. The cycle is repeated across multiple repeated layers in advanced chips—gradually constructing the complex three-dimensional structures characteristic of modern semiconductor devices.

Each layer must be aligned with exceptional accuracy, as any significant error in positioning, focus, or feature size can compromise yield or device performance. Lithography thus represents a form of microscopic choreography executed at an industrial scale.

A fundamental constraint in lithography is the wavelength of light used in the process. Shorter wavelengths enable the printing of finer features, while longer wavelengths limit the minimum achievable feature size. This physical principle has shaped the evolution of semiconductor manufacturing.

For many years, the industry utilized deep ultraviolet (DUV) lithography, employing excimer lasers at wavelengths such as 248 nanometers and later 193 nanometers. DUV tools enabled significant advancements in semiconductor scaling and facilitated key generations of chip production. However, as the industry pursued smaller nodes, DUV technology encountered limitations. Engineers introduced techniques such as multi-patterning, which divides a single pattern into multiple exposures to achieve finer structures than a single exposure would allow. While effective, multi-patterning increased process complexity, added steps, reduced throughput, and raised manufacturing costs.

Ultimately, the industry required a more transformative solution.

Extreme ultraviolet (EUV) lithography addressed this need by operating at a wavelength of 13.5 nanometers, compared to the 193-nanometer light used in previous systems. This roughly fourteen-fold reduction in wavelength enables the direct patterning of much finer features, facilitating advanced nodes such as 7 nanometers, 5 nanometers, and 3 nanometers, and supporting the development of 2-nanometer technology and beyond. EUV lithography thus revitalized the scaling frontier at a time when previous techniques had become increasingly complex.

This advancement represented a major breakthrough in modern chipmaking, though it introduced significant technical challenges.

Generating and manipulating EUV light presents substantial challenges. Unlike conventional optical systems, EUV light cannot be managed with standard lenses or materials, as most substances—including air and glass—readily absorb it. Consequently, the entire lithography process required fundamental reengineering, making EUV lithography one of the most complex engineering achievements in commercial manufacturing.

EUV light generation in ASML's systems employs a technique called laser-produced plasma. Small droplets of molten tin are injected at high speed into a vacuum chamber. An initial laser pulse strikes the droplet, shaping it for optimal interaction. A subsequent, more intense laser pulse vaporizes the droplet into plasma at temperatures exceeding those of the sun's surface. This plasma emits extreme ultraviolet light at the required wavelength of 13.5 nanometers, with the process occurring tens of thousands of times per second.

Modern semiconductor production relies on the repeated transformation of microscopic tin droplets into superheated plasma within a vacuum chamber to generate the precise wavelength of light required for atomic-scale patterning on silicon. This complex process has been successfully integrated into the global semiconductor supply chain.

In addition to generating EUV light, guiding it presents further challenges. Since EUV is absorbed by conventional optical materials, ASML's systems utilize highly specialized reflective optics composed of multilayer mirrors with exceptional smoothness and precision, called the Bragg reflectors. These mirrors are engineered to preserve the faint EUV signal as it traverses the system. The entire optical path is maintained in a high-vacuum environment to prevent absorption by air.

The resulting system resembles a controlled scientific environment rather than a conventional manufacturing tool. Each component is designed to preserve, direct, and utilize a form of light that presents significant challenges for industrial application, yet the system must operate with sufficient reliability to support the mass production of high-value semiconductor chips.

After EUV light is generated and directed through the reflective optical system, the mask pattern is projected onto the photoresist-coated wafer. The image is reduced and transferred with nanometer-level precision, while the wafer is positioned by high-speed, highly accurate stage systems. This movement requires exceptional control to ensure the pattern is consistently repeated across each die and layer. Modern EUV systems achieve resolutions that enable the fabrication of structures measuring tens of nanometers or less, allowing manufacturers to integrate billions of transistors onto a single chip.

Consequently, ASML's lithography systems are regarded as industry benchmarks.

These systems rank among the largest, heaviest, and most complex machines in advanced manufacturing. A single EUV system can weigh approximately 150 tons and cost hundreds of millions of dollars, depending on model and configuration. Each machine comprises over 100,000 parts, requires shipment in hundreds of crates, and must be assembled with meticulous care at the customer's fabrication facility. Installation may require several months to complete.

ASML's primary EUV product lines, including the TWINSCAN EXE series, are integral to the manufacturing strategies of leading semiconductor companies, including TSMC, Intel, and Samsung. This reliance grants ASML significant strategic influence, as its technology plays a decisive role in determining which firms can effectively compete at the forefront of semiconductor development.

The subsequent advancement is High-NA EUV, where 'NA' denotes numerical aperture, a critical parameter in optical resolution. High-NA EUV systems are intended to enhance imaging capabilities, enabling finer patterning and facilitating the transition to nodes below 2 nanometers. The benefits include reduced reliance on multi-patterning, improved resolution for smaller features, increased transistor density, and potentially accelerated production at future technology nodes.

However, the transition to High-NA EUV presents increased technical and financial challenges. These systems necessitate larger optics, more advanced mechanical structures, greater precision, and significantly higher capital investment. Progress at the leading edge of semiconductor technology typically involves increased complexity, fragility, and cost, yet the industry continues to advance despite these obstacles.

ASML's near-monopoly in advanced EUV lithography systems positions it as a critical player within the global semiconductor ecosystem. Without EUV technology, manufacturing the most advanced chips for artificial intelligence, cloud infrastructure, flagship smartphones, and autonomous systems would be significantly more challenging or economically unfeasible. Consequently, ASML operates at the intersection of optical engineering, industrial policy, and geopolitics.

The geopolitical implications of advanced lithography are increasingly prominent. Since access to ASML's most advanced machines strongly affects the ability to manufacture leading-edge chips, advanced lithography has become increasingly entangled export controls, alliance politics, and strategic technology policy. Control over such technological bottlenecks can extend beyond commercial interests and shape broader patterns of technological influence.

Future developments in lithography will extend beyond increased exposure precision. Anticipated advancements include higher power EUV sources to enhance wafer throughput, improved process control—potentially assisted by artificial intelligence—to increase yields and reduce variability, and the development of new photoresist materials to better accommodate EUV photon energies and complex pattern formation. Additionally, advanced packaging and three-dimensional chip stacking will increasingly complement lithography, as future performance improvements may derive from both smaller features and enhanced integration of multiple dies and functions.

Lithography remains central to semiconductor innovation, yet it now operates within a broader ecosystem. Continued progress depends not only on transistor scaling, but also on the integration of improved patterning, architecture, packaging, interconnects, and materials. Nevertheless, lithography serves as the foundational process step; without the ability to print integrated circuit patterns, architectural advancements cannot be realized.

ASML is actively developing tools aimed at the requirements of future artificial intelligence hardware and increasingly complex chip architectures. As chips become more specialized and system-level integration becomes increasingly important, the demand for precise, reliable, and high-throughput lithography will intensify. The next generation of processors, accelerators, and advanced computing systems will continue to rely on the industry's capacity to define structures at extremely small scales.

Although semiconductor progress increasingly involves chiplets, advanced packaging, and system-level design, the central role of lithography remains undiminished and is becoming more prominent in some respects. As transistor dimensions approach atomic limits, the challenge of patterning matter at such scales becomes a defining technological issue.

Advanced lithography is not merely a process step; it is one of the cornerstones of semiconductor scaling. Decades of research, engineering effort, and substantial investment have enabled ASML to develop machines that allow the industry to surpass previous physical and economic limitations.

As artificial intelligence, high-performance computing, immersive digital platforms, advanced automation, and emerging quantum and photonic systems become increasingly prevalent, lithography will remain a foundational technology supporting these advancements.

The future of semiconductors will depend on multiple factors.

However, continued progress will ultimately depend on the ability to pattern increasingly small features using advanced lithographic techniques.

At this stage in semiconductor development, the industry adopted unconventional optical methods, such as laser vaporization of tin droplets in a high-vacuum environment, to enable continued miniaturization. This innovation has become central to the semiconductor ecosystem, contributing to the emergence of a multibillion-dollar industry.

References

1. ASML, EUV Lithography Systems.
2. ASML, TWINSCAN EXE:5000.
3. ASML, 5 Things You Should Know About High-NA EUV Lithography.
4. ASML, Lenses and Mirrors - Lithography Principles.
5. ASML, 6 Crucial Steps in Semiconductor Manufacturing.
6. ASML, Making EUV: From Lab to Fab.
7. ASML, Intel and ASML Strengthen Their Collaboration to Drive High-NA into Manufacturing in 2025.
8. ASML, Our History.
9. ASML, 2025, Annual Report.
10. ASML, ASML Launches New TWINSCAN 300 mm Lithography Platform.
11. ASML Press release, ASML Launches New AT:1100 ArF System
12. Reuters, ASML's Next Chip Challenge: Rollout of Its New $350 Million High-NA EUV Machine.

13. Reuters, The $250 Million ASML Printer Behind Nvidia's Chips.
14. Reuters, ASML Unveils EUV Light Source Advance That Could Yield 50% More Chips by 2030.
15. Reuters, ASML Says Next-Gen EUV Tools Ready to Mass-Produce Chips, Marking Key Shift for AI Chip Production.
16. Light: Science & Applications, 2025, Advancements and Challenges in Inverse Lithography Technology.
17. Microsystems & Nanoengineering, 2017, Nanoimprint Lithography Steppers for Volume Fabrication of Leading-Edge Semiconductor Integrated Circuits.

Chapter 14: 3D Chips and Advanced Packaging – Building Beyond the Flat Chip

For decades, the semiconductor industry advanced through a simple and powerful formula: make transistors smaller, fit more of them onto a chip, and let the resulting gains in speed, efficiency, and cost reshape the world. That formula worked so well for so long that it came to feel almost inevitable. Moore's Law became the rhythm of modern computing, and entire industries were built on the assumption that the next generation of silicon would always be faster, denser, and more capable than the last.

However, no technological trend continues indefinitely without encountering resistance. As chipmakers advanced to nodes such as 3 nm and 2 nm, the traditional model of progress faced significant physical and economic constraints. While further transistor scaling remained technically feasible, it became increasingly challenging, costly, and complex. Consequently, the semiconductor industry began to explore alternative strategies to sustain performance improvements.

One important response has been the growing adoption of advanced packaging and three-dimensional chip integration.

This development represents a substantial shift in the industry's approach to performance. Historically, the primary focus was on maximizing functionality within a single silicon die. Increasingly, the emphasis has shifted to assembling, stacking, and interconnecting multiple silicon chips to function as a unified, more powerful system. Rather than consolidating all functions onto a monolithic die, engineers now distribute functions across multiple chips, integrating them to enhance performance, reduce power consumption, and enable more flexible designs.

In many respects, this constitutes one of the most significant changes in semiconductor architecture since the advent of the integrated circuit.

Traditional processors were often built as monolithic chips. CPU cores, cache, memory controllers, I/O logic, and other components all sat on one large slab of silicon. That approach had advantages, especially when transistor scaling was progressing smoothly. But as chips grew larger and more complex, the disadvantages became harder to ignore. Bigger dies are more vulnerable to manufacturing defects. If a defect appears on a large chip, the entire die may be lost. That hurts yield, raises cost, and makes manufacturing more punishing. Large chips also create thermal problems, as so much functionality is packed into a single area, generating more heat and complicating cooling.

Advanced packaging fundamentally alters this dynamic.

Instead of forcing everything onto a single giant piece of silicon, chip designers can divide a processor into smaller functional units, sometimes called chiplets, and then combine them into a single package. These smaller components can be manufactured separately, often with better yields and lower cost, and later linked together through extremely fast interconnect technologies. The result is a system that behaves like an integrated processor but is built from multiple pieces.

This approach offers several simultaneous advantages. Smaller dies are easier to manufacture with higher yield, and costs can be managed more efficiently. Design becomes more modular, and different system components can be fabricated using different process nodes. This enables the use of advanced and costly technologies only where they provide the greatest benefit. For example, a company may implement state-of-the-art CPU cores on a leading-edge node while utilizing mature processes for analog functions, I/O circuits, or other components that do not require advanced scaling. This strategy represents a more effective allocation of complexity, rather than merely a cost-saving measure.

Three-dimensional integration extends this concept further.

In a conventional package, components are placed side by side. In 3D integration, semiconductor dies can be stacked vertically, reducing the distance that data has to travel. Instead of signals crossing relatively long horizontal paths across a board or substrate, they can move more directly between silicon layers. That reduces latency, increases bandwidth, and lowers power consumption. In modern computing, where data movement is often as important as raw computation, those benefits are enormous.

This is one of the core reasons 3D integrations has become so attractive. In many advanced systems, especially those used for artificial intelligence and high-performance computing, the main bottleneck is not always processing capability alone. It is the ability to move data quickly and efficiently between compute units and memory. By stacking dies and shortening those paths, engineers can build systems that are faster and more energy efficient than traditional layouts would allow.

The key enabling technology behind much of this stacking is the through-silicon via, or TSV.

A TSV is a microscopic vertical electrical connection that passes through the silicon, linking one die to another in a stacked structure. Rather than routing signals around the periphery, TSVs enable power and data to move directly through silicon layers. This results in shorter electrical paths, higher transfer rates, reduced power requirements, and more compact form factors. TSVs exemplify an engineering innovation that appears straightforward in retrospect but is highly challenging to implement reliably.

TSVs have become especially important in memory.

One of the most successful applications of 3D integration is high-bandwidth memory, or HBM. HBM stacks multiple DRAM dies vertically and connects them via TSVs, creating memory systems that deliver enormous bandwidth while consuming less power than many traditional memory configurations. This is vital in applications where processors must move huge amounts of data at high speed. AI accelerators, GPUs, supercomputers, and other high-performance systems increasingly depend on HBM because conventional memory architectures cannot keep pace with the computational demands of modern workloads.

Integrating HBM in close proximity to the processor via advanced packaging significantly enhances overall system capability. The processor accesses data more rapidly, the memory subsystem operates more efficiently, and the architecture is better suited for compute-intensive applications such as machine learning training and scientific simulation. In the current era of artificial intelligence, memory bandwidth has become so critical that packaging and memory design are often as strategically significant as processor core design.

Alongside memory stacking, the rise of chiplet architecture has become a defining trend in advanced packaging.

Chiplets represent a departure from the old obsession with building one giant all-in-one die. Instead, systems are broken into specialized pieces. One chiplet may contain CPU cores. Another may handle graphics. Another may provide I/O functions. Another may include AI acceleration or memory control. These components are then connected within the package using high-speed fabrics, enabling them to behave as a coordinated whole.

This approach provides substantial flexibility. It enables faster product development, as companies can reuse or update specific chiplets without redesigning an entire monolithic processor. Fabrication risk is reduced because smaller dies are more efficiently manufactured. Different process nodes can coexist within a single product, and architectural experimentation is supported in ways that monolithic design often renders prohibitively expensive.

AMD has been one of the most visible examples of how effective this can be. Its adoption of chiplet-based designs helped build competitive, high-performance processors while managing manufacturing complexity and costs. That success did not go unnoticed. The broader industry increasingly treats chiplets not as a niche strategy, but as a major direction for future semiconductor design.

To support these new architectures, a range of advanced packaging technologies has emerged.

One notable method is 2.5D packaging, which employs a silicon interposer between the chiplets and the package substrate. The interposer serves as a high-speed communication layer, enabling multiple dies to exchange data at very high bandwidth. This technique is particularly important in GPU and AI accelerator designs, where compute dies must communicate rapidly with high-bandwidth memory and other components. The primary objective is to achieve dense integration without necessitating full vertical stacking of all logic dies.

Another significant approach is fan-out wafer-level packaging (FOWLP). This technique redistributes chip connections across a wider area, enabling thinner, smaller, and more compact packages. FOWLP is widely used in smartphones, wearables, and other consumer devices where space constraints and performance requirements are critical. Fan-out packaging allows manufacturers to reduce device thickness without sacrificing capability.

Hybrid bonding is another major advance and is widely seen as an important enabler of denser future 3D integration. Instead of relying on traditional solder bumps to join structures, hybrid bonding directly connects copper pads between wafers or dies. This enables extremely fine-pitch interconnects, lower electrical resistance, higher bandwidth, and better energy efficiency. As logic stacking becomes more advanced, hybrid bonding is likely to become a critical technology, enabling far denser, more effective connections between layers.

All of this matters profoundly in the age of artificial intelligence.

AI workloads require enormous computational power and, just as importantly, extraordinary memory bandwidth. One of the biggest bottlenecks in AI systems is data movement between processing engines and memory. Advanced packaging addresses that problem directly by enabling the combination of compute dies, stacked memory, and ultra-fast interconnects into highly optimized packages. The result is faster data movement, lower latency, and better overall system efficiency.

That is why many next-generation AI accelerators depend so heavily on packaging innovation. The chip package is no longer just a protective shell around the real technology. In advanced AI systems, the package is a central part of the architecture. It determines how closely logic and memory can be collocated, how efficiently data moves, how much heat it can handle, and how performance scales across complex computational workloads.

Another important development is the rise of the system-in-package, or SiP.

A SiP integrates multiple semiconductor dies, sensors, passive components, and sometimes other functional elements into a single compact package that functions as a complete electronic system. This approach is particularly valuable in smartphones, IoT devices, wearables, and automotive electronics, where integration, size, and power efficiency are essential. By consolidating functions into one compact unit, SiP technology enables smaller devices, increased functionality, and often lower power consumption.

Of course, none of this comes without complications.

Three-dimensional integration and advanced packaging introduce serious engineering challenges. Thermal management is one of the biggest. When dies are stacked closely together, heat can become trapped inside the package, creating reliability risks and performance constraints. Cooling a flat chip is hard enough. Cooling a stack of them is worse. Engineers must carefully design thermal pathways to remove heat effectively without compromising long-term performance.

Manufacturing complexity is another significant challenge. Stacking, bonding, aligning, and testing multiple dies with extremely fine tolerances is highly demanding. Yield management becomes more complicated, as defects in one layer can impact the entire package. Reliability testing and inspection have also become more rigorous. Thus, while advanced packaging addresses certain issues, it introduces new engineering challenges.

To address these physical and thermal limitations, the semiconductor industry is advancing toward a new technological frontier defined by fundamental changes in materials, power delivery, and optical communication.

A significant recent advancement is the Backside Power Delivery Network (BSPDN). In conventional monolithic designs, both data signals and power are routed from the front side of the silicon, which leads to substantial routing congestion and thermal hotspots as chips are stacked. Relocating the entire power grid to the wafer's underside enables separating power delivery from data transmission. This reorganization of the silicon substantially reduces voltage drop, minimizes interference, and allows transistors to operate at higher speeds and lower temperatures, which is essential for the latest semiconductor nodes. In early 2026, backside power delivery crossed from research into commercial reality, with Intel leading the way. Its PowerVia technology, integrated into the Intel 18A node, became the first backside power delivery network to enter high-volume manufacturing and ship into consumer devices. TSMC is expected to release its Super Power Rail technology on the A16 (1.6nm) node.

Concurrently, the physical foundation of advanced packaging is undergoing significant transformation. As chiplet architectures increase in size and complexity, traditional organic substrates are approaching their performance limits and frequently warp under the intense heat generated by compute-intensive workloads. As a result, leading manufacturers are interested in adopting glass substrates, which provide superior rigidity and dimensional stability, maintaining package flatness and a thermal expansion profile that matches silicon. Additionally, glass enables much denser electrical and optical routing. This transition facilitates the integration of more chiplets and high-bandwidth memory stacks within a single package, without the electrical noise or physical instability associated with older materials.

As package density and speed increase, traditional copper wiring encounters a fundamental limitation: it cannot transmit data rapidly across server racks without producing excessive heat. This challenge has increased industry interest in Co-Packaged Optics (CPO). Rather than converting electrical signals to light at the edge of the circuit board, CPO integrates fiber-optic transceivers directly into the chip package. This approach has moved beyond architectural theory into early commercial deployment, with Broadcom's Bailly first-generation 51.2T switch standing out as a prominent example. This shift allows large AI clusters to scale beyond the thermal limits of copper, enabling communication through photons from the source and delivering the bandwidth needed for the next generation of compute infrastructure.

Still, the broader trajectory is clear. As conventional transistor scaling slows, packaging is becoming one of the major drivers of semiconductor progress.

The industry is moving toward heterogeneous integration, in which chips built with different technologies, materials, and process nodes are combined into one system. Future directions are likely to include wafer-to-wafer stacking for high-volume production, optical interconnects within packages, AI-assisted optimization of chiplet-based systems, and more sophisticated cooling methods tailored to dense 3D structures. The performance gains of tomorrow may come less from shrinking one transistor by another tiny amount and more from intelligently assembling multiple specialized components into highly efficient systems.

This represents a fundamental shift in perspective.

For most of the semiconductor era, progress meant smaller devices on a single die. Increasingly, progress now means smarter assembly, denser integration, and better coordination among multiple silicon chips. The future of computing may depend as much on how chips are packaged and interconnected as on how small their transistors become.

This is why 3D chips and advanced packaging matter so much. They represent one of the most significant redefinitions of semiconductor design since the integrated circuit itself. By enabling stacking, modular chiplets, high-speed interconnects, and heterogeneous integration, advanced packaging offers the industry new ways to continue improving performance even as traditional scaling becomes more difficult.

This stage in semiconductor history marks the transition from a focus on reducing chip size to one on stacking and integrating components with careful attention to thermal management. This evolution reflects both the complexity and ingenuity of modern semiconductor engineering.

The next era of semiconductor innovation will not be built solely on smaller transistors. It will also depend on how intelligently those transistors are assembled into systems.

References

1. TSMC, 3DFabric.
2. TSMC, CoWoS.
3. TSMC, TSMC-SoIC.
4. TSMC, Integrated Fan-Out Wafer Level Packaging.
5. AMD, AMD Chiplet Ecosystem.
6. AMD, AMD CDNA 4 Architecture.
7. AMD, Introducing the AMD Instinct MI300 Series Accelerators.
8. AMD, Instinct MI300A Data Sheet.
9. Intel Foundry, Advanced Packaging Innovations.
10. Intel, Foveros Direct 3D Technology Brief.
11. Intel, Embedded Multi-die Interconnect Bridge Product Brief.
12. Intel, Quality and Reliability Vol. 6: Intel Advanced Packaging Technology.

13. Micron, HBM2E: Our Fastest Memory for the Modern Data Center.
14. Micron, HBM3E: Powering the Future of AI with High-Bandwidth Memory.
15. Intel, 2023, Intel PowerVia: Industry-First Backside Power Delivery.
16. TSMC, 2024, TSMC Introduces A16 Technology, Bringing Backside Power Rail to the Angstrom Era.
17. Intel, 2023, Intel Leads the Industry with Glass Substrates for Advanced Packaging.
18. Samsung Electronics, 2024, Samsung Outlines 2026 Roadmap for Glass Substrate Commercialization at CES 2024.
19. Broadcom, 2024, Broadcom Ships Bailly: The World's First 51.2 Tbps Co-Packaged Optics Ethernet Switch.
20. Broadcom, 2025, Third-Generation Co-Packaged Optics (CPO) Technology with 200G/lane Capability.
21. Optical Internetworking Forum, 2024, Implementation Agreement for Co-Packaged Optics 3.2T Module.

PART IV: New Materials and Radical Computing

Chapter 15: Beyond Silicon – The Search for New Foundations of Computing

For over fifty years, silicon has been the foundational material of the semiconductor industry. Few materials have influenced modern civilization as profoundly while remaining largely unnoticed by the general public. Silicon enabled the practical development of integrated circuits and facilitated technological scaling. Its combination of abundance, stability, and manufacturability coincided with the rise of electronics as a global industry. From early microprocessors to advanced AI accelerators, silicon has served as the essential substrate of the digital era.

Silicon's dominance resulted from its practical advantages at scale. It could be purified, fabricated into wafers, patterned with high precision, and integrated into a manufacturing ecosystem that evolved into a highly sophisticated industrial system. In the development of semiconductors, silicon proved to be not only the most practical mainstream material for large-scale semiconductor manufacturing but also the foundation upon which the modern engineering infrastructure was established.

However, even the most dominant materials encounter limitations.

As transistors approach atomic dimensions and computing demands intensify, silicon is exhibiting significant limitations. The issue is not a sudden loss of utility, but rather that the requirements for semiconductors have become more diverse and demanding. High-performance processors generate increased heat, power systems require greater efficiency, communications infrastructure demands faster switching at higher frequencies, and electric vehicles necessitate devices capable of withstanding high voltages and challenging thermal conditions. Advanced computing seeks higher speeds with reduced energy consumption. While silicon remains highly capable, it can no longer address all requirements with equal effectiveness.

Consequently, the semiconductor industry is exploring materials beyond silicon, not as replacements, but as supplements. It is improbable that a single material will entirely supplant silicon. Instead, the future will likely involve a diverse set of materials, each selected for optimal performance in specific applications. Among these, gallium nitride, silicon carbide, and graphene are particularly significant.

Each of these materials represents a distinct advancement in semiconductor technology.

Understanding the significance of these alternative materials requires examining the limitations of silicon. For decades, silicon transistors advanced under Moore's Law, with continual reductions in size, increased density, lower costs, and enhanced performance. However, as feature sizes have reached the nanometer scale, challenges such as increased leakage current, greater heat generation, and diminishing efficiency gains have become prominent. In many high-voltage, high-frequency, or high-temperature applications, wide band-gap materials can offer performance advantages over silicon.

These limitations are significant because the semiconductor industry now serves a wide range of applications beyond conventional computing. Electric vehicles require efficient power conversion, renewable energy systems demand durable high-voltage devices, and advanced communications networks such as 5G need fast, efficient radio-frequency components. Data centers seek improved power delivery and reduced energy losses. In these contexts, the primary concerns are energy control, rapid switching, and resilience under demanding conditions, rather than simply increasing transistor density. This shift underscores the importance of new materials.

Many of the leading alternative materials are classified as wide bandgap semiconductors. These materials possess a larger energy gap between the valence and conduction bands compared to silicon, resulting in advantageous properties for power and high-frequency electronics. Wide bandgap semiconductors can withstand higher voltages, operate at elevated temperatures, and switch more rapidly with reduced energy losses. These characteristics enable engineers to address challenges that are difficult or inefficient to solve using conventional silicon.

Currently, gallium nitride and silicon carbide are among the most commercially significant wide bandgap materials.

Gallium nitride (GaN) has become a prominent material in modern electronics. Initially recognized for its applications in optoelectronics and LEDs, GaN's significance has expanded to power electronics and communications. It exhibits high electron mobility, supporting high-speed operation, performs effectively at high frequencies, and enables efficient power conversion while handling higher voltages than many traditional silicon devices. These properties make GaN particularly valuable in applications requiring both speed and efficiency. GaN is increasingly utilized in 5G base stations, radar systems, satellite communications, compact power supplies, and fast-charging devices. For consumers, modern fast chargers exemplify GaN's advantages, as they can be manufactured to be smaller, lighter, and more efficient than traditional silicon-based chargers while maintaining high power output. This demonstrates how advanced materials can alter engineering trade-offs.

Similar principles apply at larger scales in data centers and communications systems. Improved power conversion efficiency reduces heat generation, enables more compact hardware, and enhances overall system performance. In sectors with extensive device deployment, even modest efficiency improvements are strategically significant. As a result, GaN is not only of scientific interest but is also actively transforming products and infrastructure.

Silicon carbide (SiC) is another material of increasing strategic importance, particularly in high-power applications. While GaN excels in high-frequency and compact power systems, SiC is especially valuable in high-voltage and high-temperature environments. Compared to silicon, SiC can withstand significantly higher voltages, switch with lower energy losses, conduct heat more efficiently, and operate at temperatures unsuitable for conventional silicon.

These properties make SiC particularly suitable for electric vehicles, renewable energy systems, industrial power equipment, and charging infrastructure. As a result, SiC has become a key material in the electrification sector. Electric vehicles depend on efficient power electronics to manage energy transfer between batteries, motors, and charging systems. Enhanced efficiency in power electronics reduces energy loss as heat, improves system performance, and can increase vehicle range, which is critical in the competitive electric vehicle market.

Consequently, major automotive manufacturers and suppliers are increasingly adopting SiC-based devices for inverters, onboard chargers, and power-control systems. Charging infrastructure also benefits from higher efficiency and improved thermal performance, enabling faster and more reliable power delivery. As electrification advances, silicon carbide is anticipated to play an even more significant role. SiC exemplifies how new semiconductor materials are practical solutions already transforming key industries.

Beyond GaN and SiC, the semiconductor industry is increasingly focusing on ultra-wide-bandgap materials designed for even more extreme operating conditions. Gallium oxide (Ga2O3) is widely considered a promising next step for high-voltage power electronics because its significantly larger band gap compared with SiC and GaN could enable smaller, more efficient, and higher-voltage power converters. Diamond, on the other hand, is often seen as the ultimate long-term semiconductor material due to its exceptional thermal conductivity and very high breakdown strength. In addition to these properties, diamond's exceptional thermal stability enables it to retain semi-conducting capabilities at temperatures above 500°C, conditions under which traditional silicon would fail entirely. Although its practical use is still limited by high cost and the technical challenge of controlled doping, diamond remains the most ambitious goal in the search for advanced power semiconductor materials.

Graphene is another notable, though technically challenging, emerging semiconductor material. Graphene consists of a single layer of carbon atoms arranged in a hexagonal lattice. It is recognized as one of the thinnest, strongest, and most conductive materials known. Electrons move through graphene at exceptional speeds, and it exhibits outstanding thermal conductivity and mechanical strength despite its thinness. These properties make graphene a highly promising material for some future electronic and sensing applications,

However, graphene presents a significant limitation for digital electronics: it lacks a natural bandgap. This limitation is significant because digital transistors require the ability to reliably switch off current flow. Without a bandgap, graphene cannot function effectively as a digital switch, making it unsuitable for conventional transistor logic, where reliable on-off behavior is essential.

Researchers are overcoming this limitation with techniques like graphene nanoribbons, bilayer graphene structures, and hybrid systems combining graphene with other materials. These methods aim to create effective band gaps or harness graphene's exceptional transport properties while maintaining control over switching behavior. Although some progress has been made, graphene is still less established than GaN or SiC in mainstream semiconductor applications.

GaN enables the development of smaller chargers, faster communications systems, and more efficient power conversion. SiC contributes to increased electric vehicle range, more efficient charging infrastructure, and reduced energy waste in industrial systems. If graphene's switching limitations can be addressed, it may facilitate ultra-fast electronic systems and novel device classes. These advancements have significant implications for transportation, energy, computing, communications, and infrastructure.

Graphene offers great potential but also poses significant engineering challenges, prompting researchers to explore other two-dimensional materials, such as Transition Metal Dichalcogenides (TMDs), such as Molybdenum Disulfide (MoS_2). Unlike graphene, these materials are atomically thin but can be turned off efficiently. They provide a possible way to keep shrinking transistors down to the single-atom level, effectively extending Moore's Law through a molecular approach to computing.

Accordingly, the industry is shifting from material uniformity to a strategy based on the optimal use of diverse materials.

Functional requirements will increasingly be matched to the specific physical strengths of various materials. Future systems may integrate silicon logic, GaN radio-frequency components, SiC power-control devices, and potentially new materials for sensing, photonics, or advanced interconnects. While the semiconductor industry has traditionally focused on functional integration, leading towards heterogeneous integration of a broader range of materials.

The rise of photonic integrated circuits, built on materials like indium phosphide and lithium niobate, shifts data movement from electrons to photons, reducing heat and resistance and providing a powerful solution to the bandwidth and energy bottlenecks facing AI data centers. Meanwhile, neuromorphic computing encourages a deeper rethinking of computation, using memristive materials such as hafnium oxide and perovskites to integrate memory and processing in a single device. This challenges the energy-intensive separation of processor and memory typical of conventional silicon systems and opens the door to more efficient, brain-inspired computing. Together, these emerging trends suggest that the future of semiconductors will be shaped not only by improved silicon but also by new material platforms that expand the possibilities of computing.

However, realizing this future presents significant challenges. Silicon continues to dominate due to the extensive optimization of global manufacturing systems for its production. Wafers, tools, process flows, defect control, supply chains, and engineering practices have been refined over decades for silicon-based manufacturing. Integrating new materials into this ecosystem is complex and costly. Wafer production becomes more difficult, defect densities may increase, and equipment often requires adaptation. Achieving economic manufacturing scale is a distinct challenge beyond demonstrating scientific feasibility in laboratory settings.

For this reason, the transition beyond silicon is expected to be gradual. The primary barriers are industrial rather than purely scientific. A material may possess excellent properties but still fail to achieve widespread adoption if it cannot be manufactured reliably, affordably, and at scale. The semiconductor industry contains many examples of promising innovations that did not achieve commercial success due to these challenges.

Nevertheless, pursuing these advancements is justified by the substantial potential benefits.

The future beyond silicon does not represent a rejection of silicon, but rather an expansion of available semiconductor materials. Silicon will remain central for the foreseeable future, as steel has despite the introduction of aluminum, composites, and carbon fiber. However, the diverse demands of modern technology require a range of materials, and progress increasingly depends on selecting the most appropriate material for each application.

This transition marks an especially dynamic period for the industry. Future semiconductor innovation will rely not only on advances in transistor scaling or packaging, but also on the effective utilization of new materials with unique properties and industrial potential. Breakthroughs are likely to result from leveraging the strengths of alternative materials rather than relying solely on silicon. This development signifies the emergence of a new frontier in semiconductor materials.

Gallium nitride, silicon carbide, and graphene are among the leading materials driving this new frontier. Collectively, they are expanding the capabilities of semiconductor technology in areas such as faster communications, more efficient power systems, enhanced electrification, and potentially new computing paradigms.

The horizon of the beyond-silicon era extends well beyond traditional rigid, high-power systems. A concurrent transformation is occurring through the development of organic semiconductors, where carbon-based polymers are enabling a new generation of flexible, lightweight, and printable electronics. In contrast to the brittle wafers used in conventional semiconductor manufacturing, these materials can be deposited onto substrates such as plastic, glass, or fabric, enabling the creation of wearable health monitors, flexible displays, and a broader array of bioelectronic devices that integrate seamlessly with daily life.

Simultaneously, the pursuit of significantly greater energy efficiency has focused attention on quantum materials, including topological insulators, which have the potential to transport information with minimal resistive loss. By leveraging the unique properties of quantum mechanics, these materials may bridge the gap between current power-intensive processors and future ultra-efficient architectures. Considered alongside the light-harvesting capabilities of perovskites, the post-silicon landscape is characterized not only by increased speed and power but also by enhanced adaptability, energy efficiency, and integration into the physical fabric of modern society.

At this stage, silicon is no longer the sole focus of semiconductor innovation. It remains the foundation, but the future will belong to a broader material ecosystem shaped by the uncompromising demands of physics and the expanding ambitions of human technology.

References

1. Infineon, Gallium Nitride.
2. Infineon, GaN Transistors.
3. Infineon, CoolGaN Technologies.
4. Texas Instruments, Gallium Nitride Power Stages.
5. Texas Instruments, GaN: Pushing the Limits of Power Density and Efficiency.
6. Wolfspeed, Silicon Carbide Power Products.

7. Wolfspeed, Introducing the Industry's First Commercially Available 10,000V Silicon Carbide Power MOSFET.
8. onsemi, Silicon Carbide Technology.
9. onsemi, How SiC Is Shifting the Future of EV.
10. National Renewable Energy Laboratory, Wide Bandgap Technology Enhances Performance of Electric-Drive Vehicles.
11. National Renewable Energy Laboratory, Wide-Bandgap Electronics Analysis.
12. APS, Random Walk to Graphene.
13. IBM Research, Graphene Field-Effect Transistors with High On Off Current Ratio and Large Transport Band Gap at Room Temperature.
14. PMC, Graphene Properties, Synthesis and Applications: A Review.
15. PMC, Open-Bandgap Graphene-Based Field-Effect Transistor.
16. Nano Letters, 2024, Ultrawide Bandgap Diamond ε-Ga2O3 Heterojunction pn Diodes with Breakdown Voltages over 3 kV.
17. Nano Letters, 2025, Sub-Nanometer Equivalent Oxide Thickness and Threshold Voltage Control Enabled by Silicon Seed Layer on Monolayer MoS2 Transistors.
18. Nature Photonics, 2025, Non-Hermitian Hybrid Silicon Photonic Switching.
19. Nature Communications, 2026, High power density gallium nitride radio frequency transistors via enhanced nucleation in heteroepitaxy.
20. Nature Materials, 2020, Organic mixed ionic–electronic conductors.
21. Nature Materials, 2026, Crystal-facet-directed all-vacuum-deposited perovskite solar cells.

22. Nature Reviews Materials, 2019, Metal halide perovskite nanostructures for optoelectronic applications and the study of physical properties.
23. Science Advances, 2026, Programmable digital printing of organic single crystals for integrated electronics.
24. ACS Energy Letters, 2026, Harnessing quantum computing for energy materials: Opportunities and challenges.
25. Nanjing University, 2025, Scaling MoS2 Transistors to 1 nm Node.
26. ACS Applied Electronic Materials, 2025, From Oxides to 2D Materials: Advancing Memristor Technologies for Energy-Efficient Neuromorphic Computing.
27. University of Southern California, 2026, USC Scientists Build a Memory Chip That Survives Temperatures Hotter Than Lava.

Chapter 16: Quantum Computing – The Next Frontier of Computing Power

For more than seventy years, classical computers have reshaped human civilization. Built first from vacuum tubes, then transistors, then integrated circuits of astonishing density, they have carried humanity from room-sized machines to smartphones, cloud infrastructure, artificial intelligence platforms, and global digital networks. Their rise has been one of the great technological transformations in history, driven largely by the steady improvement of semiconductor hardware and the relentless logic of scaling.

Despite their significant capabilities, classical computers encounter fundamental limitations.

Some problems become so complex that even the most advanced supercomputers struggle under the sheer weight of the calculations required. Simulating the behavior of molecules, modeling complex materials, breaking certain cryptographic schemes, or solving massive optimization problems can demand computing resources that grow explosively as the problem gets larger. In these domains, the challenge is not simply that classical computers are too slow today. It is that the problem's structure itself fights back, punishing brute-force approaches with exponential complexity. Quantum computing may offer advantages for some of these limitations in specific problem classes.

Quantum computing is not just a faster version of ordinary computing. It is a fundamentally different model for processing information, built not on the rules of classical electronics but on the strange, often deeply counterintuitive principles of quantum mechanics. Instead of relying on bits that exist as either 0 or 1, quantum computers use qubits, units of information that can behave in ways classical bits never can. This change in the basic unit of computation opens the door to quantum states and correlations that, in some cases, can be exploited computationally to make certain classes of problems far more tractable.

Consequently, governments, research institutions, and major technology firms such as IBM and Google have made substantial investments in quantum computing. Their motivation is not driven by trends or superficial appeal, but by the potential for practical quantum systems to address problems that remain intractable for even the most advanced classical computers.

Understanding the distinctiveness of quantum computing requires first examining the concept of the classical bit.

In classical computing, information is represented in binary form. A bit can be 0 or 1. Every piece of data, every instruction, every image, every spreadsheet, every video stream, every machine-learning model eventually rests on vast combinations of these two states. The entire digital world, for all its complexity, is built on this simple foundation.

Quantum computing is based on a fundamentally different unit of information: the qubit.

A qubit can exist in a state corresponding to 0, a state corresponding to 1, or, more strangely, in a combination of both at the same time. This is known as superposition. It does not mean the qubit is vaguely undecided in some sloppy human sense. It means the mathematics describing its state contains multiple possibilities simultaneously, and those possibilities can be manipulated before measurement collapses them into a definite outcome.

This property is a fundamental source of quantum computational power. The ability of qubits to exist in superposed states enables quantum systems to represent multiple configurations simultaneously. As the number of qubits increases, the size of the representable state space expands exponentially. Consequently, quantum computers can explore certain computational problems in ways that classical machines cannot efficiently replicate. The primary value of quantum computing lies in leveraging this unique structure to achieve advantages in specific problem domains, rather than replacing classical computers entirely.

Superposition is only part of the story. Two other central ideas are entanglement and interference.

When qubits become entangled, their states become correlated in a way that has no true classical equivalent. The behavior of one qubit becomes linked to the behavior of another, even when the individual states cannot be described independently. This does not mean information is magically teleported in the science-fiction sense, but it does mean that the system as a whole carries relationship that are deeper than the sum of its parts. Entanglement enables quantum computers to coordinate complex computational structures across multiple qubits and is one of the reasons some quantum algorithms may achieve striking performance advantages on specialized problems.

Interference is a mechanism that leverages superposition and entanglement to cancel out incorrect answers while amplifying the correct ones.

Superposition, entanglement, and interference together constitute the foundation of quantum computing's unique capabilities. These phenomena enable quantum systems to manipulate probability amplitudes and correlated states in ways that classical systems cannot replicate. However, practical utility requires systematic manipulation of these states, which is achieved through the use of quantum gates and circuits.

In a classical processor, logic gates manipulate bits. AND, OR, NOT, NAND, and similar operations combine to form the circuits that carry out computation. Quantum computers have their own version of this idea. Quantum gates operate on qubits, changing their states through carefully controlled transformations. Some gates rotate qubit states. Some create entanglement between qubits. Some prepare the system for measurement. A sequence of these operations forms a quantum circuit, which acts as the computational program.

The design of quantum circuits presents significant challenges. Quantum systems are inherently delicate, and algorithms must be carefully constructed to exploit quantum phenomena without undermining the fragile coherence essential for computation. As a result, practical quantum algorithm development remains a highly demanding discipline, characterized by complex mathematics, unstable hardware, and minimal tolerance for error.

Among the many hardware approaches being explored, one of the most developed today uses superconducting qubits.

Superconducting qubits are built from microscopic circuits that behave quantum mechanically when cooled to extraordinarily low temperatures. These devices are carefully engineered circuits whose electrical behavior enters a quantum regime under extreme conditions. They are carefully engineered circuits whose electrical behavior enters a quantum regime under extreme conditions. To maintain that regime, they are housed in dilution refrigerators that cool them to temperatures near absolute zero, colder than outer space. This is necessary because heat and environmental noise can easily destroy fragile quantum states.

IBM and Google have both made substantial investments in superconducting qubit systems. Their processors are housed within elaborate cryogenic structures, equipped with multiple layers of shielding, control electronics, and measurement hardware. Despite their unconventional appearance, these systems represent a rigorous effort to render quantum information controllable and practically useful.

The race to build better quantum hardware has become one of the most closely watched competitions in advanced computing.

IBM has publicly pursued a long-term roadmap centered on scaling hardware increasing qubit counts, improving stability, and building toward fault-tolerant quantum systems. It has also made quantum computing broadly accessible via cloud platforms, enabling researchers, developers, and students worldwide to experiment with real quantum processors and simulators. This has helped create an ecosystem rather than just a laboratory project. IBM's strategy reflects a belief that quantum computing will mature through steady scaling, broad engagement, and persistent engineering improvements.

Google, meanwhile, drew global attention in 2019 when it announced that its Sycamore quantum processor had achieved what it called quantum supremacy. The phrase referred to a milestone in which a quantum computer performed a specific task that would be impractical for a classical computer to complete in the same way. The task itself was highly specialized and not of immediate everyday use, which led to considerable debate and public confusion. Still, the result was significant because it demonstrated that quantum hardware could, under the right conditions, perform certain calculations in a way that classical systems found extraordinarily difficult to match.

This did not mean quantum computers had suddenly become superior to classical ones in general, far from it. Classical computers remain vastly more practical, reliable, and useful for most workloads. But the demonstration mattered because it suggested that quantum advantage was not purely theoretical. It could emerge in real hardware, even if only in narrow settings for now.

The specificity of these applications is a critical consideration.

Quantum computers are not expected to replace laptops, servers, or classical data centers in the foreseeable future. Their value lies in targeted applications where quantum behavior can be leveraged computationally. One of the most discussed areas is chemistry and materials science. Molecules and materials are themselves quantum systems, and simulating them accurately on classical computers is often painfully difficult. Quantum computers may eventually model molecular interactions with far greater fidelity, helping accelerate drug discovery, catalyst design, battery chemistry, and the development of new materials.

Cryptography is another major area of attention. Much of today's public-key encryption relies on mathematical problems that are difficult for classical computers, such as factoring very large numbers. Quantum algorithms, most famously Shor's algorithm, suggest that sufficiently powerful quantum computers could solve some of these problems much more efficiently. That possibility has accelerated work on post-quantum cryptography as governments, security agencies, and technology firms are interested to attempt to prepare encryption systems that can remain secure even in a future with powerful quantum machines.

Optimization is yet another area where quantum computing may prove valuable. Supply chains, logistics, traffic networks, financial portfolios, and energy distribution systems often involve enormous combinatorial spaces, making it extremely hard to find the best solution. Quantum approaches may not magically solve every such problem, but they may provide advantages in certain classes of optimization or simulation tasks, where classical methods strain under complexity.

While the potential of quantum computing is considerable, the associated challenges are equally significant.

Quantum computing remains one of the most difficult engineering challenges in modern science. One of the biggest problems is decoherence. Quantum states are fragile. Interactions with the surrounding environment, heat, noise, electromagnetic disturbance, or imperfect control can cause qubits to lose their quantum character. When that happens, the information encoded in the quantum state is degraded or destroyed. Maintaining coherence long enough to perform useful computation is therefore one of the central battles in quantum engineering.

Error rates are another major issue. Quantum operations are not perfect, and even small inaccuracies can accumulate rapidly in larger circuits. Building meaningful systems will require quantum error correction, which in turn requires many physical qubits to create a smaller number of stable logical qubits. This makes scaling especially difficult. It is not enough to build a processor with more qubits. Those qubits must be controllable, low-error, entangled reliably, and capable of supporting robust correction schemes. The road from a noisy experimental device to a large-scale, fault-tolerant quantum computer is long and steep.

For this reason, many experts anticipate that the near-term future will heavily rely on quantum-classical hybrid systems rather than standalone quantum computers.

In such systems, classical processors would handle ordinary computing tasks, orchestration, and data management. In contrast, quantum processors would be used selectively for the parts of a problem where they offer an advantage. This hybrid model is much more realistic in the medium term. It reflects a practical understanding of quantum computing as a specialized extension of the computing landscape rather than a wholesale replacement for everything that came before.

This pattern aligns with historical trends in computing, where new architectures typically complement rather than immediately replace existing ones. They tend to dominate specific domains where their strengths are most pronounced and gradually influence the broader system. Quantum computers are likely to establish their roles in scientific computing, optimization, and advanced simulation prior to becoming widely generalized technologies.

From this perspective, the current state of quantum computing is analogous to the early development of classical computing in the 1940s and 1950s. The machines remain large, complex, delicate, and challenging to scale. Practical applications are still emerging, and the broader societal impact has yet to be realized. Nevertheless, the rapid pace of research and investment indicates that the field is advancing beyond mere scientific curiosity.

Governments, universities, and technology companies recognize the significance of quantum computing. The field now has scientific, industrial, and strategic significance beyond academic research alone. In this context, quantum computing is beginning to assume a strategic importance comparable to that of semiconductors in previous decades.

If large-scale, fault-tolerant quantum systems can eventually be built, the consequences could be profound. Science could accelerate. Drug development could improve. Materials engineering could enter a new era. Energy systems could be modeled more effectively. Some current public-key cryptographic assumptions could become vulnerable if large-scale fault-tolerant quantum systems are built. Even artificial intelligence may eventually find new relationships with quantum methods in highly specialized settings.

This future has not yet materialized. Practical, large-scale fault-tolerant quantum computers remain a long-term goal, with timelines still uncertain. The engineering challenges are substantial and costly. Nevertheless, the trajectory is evident: quantum computing has transitioned from a speculative concept to an active area of industrial, scientific, and geopolitical importance.

These factors collectively underscore the importance of quantum computing.

Quantum computing constitutes one of the most ambitious frontiers in twenty-first century technology. It challenges researchers to transcend classical computation and develop machines that exploit the fundamental structure of physical reality. This pursuit is both highly ambitious and likely more complex than even its proponents acknowledge.

If successful, quantum computing could represent the next significant advancement in information processing, problem-solving, and the understanding of the physical world.

The semiconductor revolution taught humanity how to build extraordinary machines from silicon.

Quantum computing may enable the construction of future computational systems based directly on the fundamental laws of nature.

References

1. IBM Quantum, What Is Quantum Computing.

2. IBM Quantum, Hardware.
3. IBM, Quantum Roadmap.
4. IBM, Quantum Roadmap Update Through 2033+.
5. IBM Quantum Blog, IBM Cools Down World's Largest Quantum-Ready Cryostat.
6. Google Quantum AI, Official Site.
7. Google Research, Quantum Supremacy Using a Programmable Superconducting Processor.
8. NIST, Post-Quantum Cryptography.
9. NIST, NIST Releases First 3 Finalized Post-Quantum Encryption Standards.
10. NIST, PQC Standardization Process.
11. IBM Quantum, Quantum Computing and Chemistry.
12. IBM Quantum, Quantum Computing and Optimization.
13. Google Quantum AI, Research Materials on Superconducting Qubits.
14. IBM Quantum Experience, Cloud Quantum Access Materials.
15. Fermilab, It's Colossal: Creating the World's Largest Dilution Refrigerator.
16. ULVAC, partnering with IBM to Help Advance Research and Development for Quantum Computing by Jointly Developing a Next-Generation Dilution Refrigerator.
17. IBM, Fault Tolerance Roadmap Materials.
18. Google Quantum AI, Willow Materials.
19. IBM, Cryostats and Superconducting Systems Materials.
20. Google Quantum AI and IBM Quantum, Official Platforms.
21. Nature, 2019, Quantum Supremacy Using a Programmable Superconducting Processor.
22. Nature Electronics, 2022, Qubits made by advanced semiconductor manufacturing.

Chapter 17: Photonic Chips – Computing with Light

For decades, the computing world has relied on one basic physical fact: electrons can be pushed through semiconductor circuits, controlled with astonishing precision, and turned into logic, memory, and communication. That approach built the modern digital age. It gave us computers, smartphones, cloud infrastructure, artificial intelligence platforms, and the vast electronic systems that now run economies, militaries, and daily life. The rise of silicon electronics remains one of humanity's greatest engineering achievements.

However, even the most significant technological achievements encounter bottlenecks over time.

As computing systems become more powerful, one problem has grown increasingly difficult to ignore. It is no longer enough to make processors faster if the data they need cannot move quickly and efficiently enough to keep up. Modern chips contain billions of transistors, and the systems built around them shuffle staggering amounts of information between logic cores, memory, accelerators, network devices, and storage. That movement consumes energy, generates heat, and creates delays. In many advanced systems, especially those used for artificial intelligence and high-performance computing, the real constraint is not just computation. It is the cost of moving data from one place to another.

At this juncture, photonic technology becomes increasingly relevant.

Photonic chips, sometimes called optical chips, use light rather than electricity to carry information. Instead of relying entirely on electrons moving through metal interconnects and semiconductor pathways, these systems manipulate photons, the particles of light, to transmit and, in some cases, even process data. Because light travels extremely fast and dissipates far less heat than conventional electrical signals, photonic technologies offer a possible route past some of the biggest limits facing modern computing.

Photonic technology offers several important advantages in bandwidth-intensive systems. While electrons are effective for many tasks, they are not optimal for transferring large volumes of data at increasing speeds within dense, power-intensive systems. Electrons encounter resistance, generate heat, and experience signal degradation over distance. As systems scale, crosstalk and bandwidth limitations become increasingly challenging to manage. In advanced processors and servers, these issues collectively contribute to what engineers refer to as the data movement bottleneck. Thus, even systems with exceptional computational capabilities are ultimately constrained by the efficiency of information flow.

This challenge is a primary reason why photonics has transitioned from an academic curiosity to a central focus within engineering research and development.

A photonic chip is a semiconductor device that generates, guides, modulates, and detects light within an integrated circuit. Instead of wires carrying current from one block to another, optical waveguides carry light signals through carefully structured paths on the chip. These waveguides serve a role somewhat analogous to electrical interconnects, but with very different physical behavior. Alongside them sit other optical components such as lasers, modulators, photodetectors, switches, and couplers, all working together to encode information onto light and move it through a system.

Effectively, photonic chips create miniature optical networks within computing hardware. A key advantage of these networks is Wavelength Division Multiplexing (WDM). In contrast to electrons moving through copper wires, multiple wavelengths of light can simultaneously traverse a single waveguide without interference. This feature enables extensive parallelization, supporting Terabit-per-second transmission rates on a single optical pathway, a level that cannot be efficiently achieved with conventional electrical interconnects.

This development significantly expands the range of possibilities for computing systems. Light can transmit vast quantities of data at extremely high speeds and is less susceptible to resistive heating than electrical connections. Additionally, optical signals can travel longer distances with reduced loss and interference. These characteristics make photonics particularly advantageous in areas where modern electronics face substantial limitations.

One of the most important developments in this field is silicon photonics.

Silicon photonics seeks to integrate optical components directly onto silicon-based platforms using semiconductor manufacturing methods compatible, at least in part, with the broader chip industry. This matters because photonics will never become industrially important if it remains trapped in exotic laboratory processes disconnected from real manufacturing.

Silicon photonics offers a bridge between the mature world of semiconductor fabrication and the emerging world of optical data movement. By building photonic structures on silicon, engineers can leverage existing manufacturing knowledge, production infrastructure, and integration pathways while extending systems into the optical domain.

This integration offers substantial practical benefits.

Silicon photonics enables high-speed communication with enhanced energy efficiency, leveraging the scalability and established processes of conventional semiconductor manufacturing. This approach facilitates closer integration of optical and electronic functions, rather than maintaining them as separate technologies combined in a less efficient manner. In an industry that prioritizes integration, this represents a significant advantage.

The impact of photonics is already visible in data center interconnect and networking.

Modern data centers are the industrial engines of cloud computing, AI training, video streaming, online services, and global internet infrastructure. They move colossal volumes of data between servers, switches, accelerators, and storage systems. Optical communication is already widely used in these environments over longer distances, such as between racks or across broader network links. But the scale of modern workloads, especially AI workloads, is pushing photonics deeper into the system.

As server architectures become denser and more bandwidth-hungry, optical links are increasingly attractive not only for data-center networking but for communication inside the computing infrastructure itself. Future systems may rely on photonic chips to move data between processors, between accelerators, between memory pools, and across chip boundaries that are now handled electrically. This is particularly important in AI, where data movement has become a dominant performance and energy challenge.

Artificial intelligence systems are especially punishing in this regard. Training large models requires a large number of processors and accelerators working together, constantly exchanging parameters, activations, gradients, and memory traffic. Conventional electrical interconnects struggle under the pressure. They consume too much power, create too much heat, and become more problematic as system scale increases. Optical interconnects offer a compelling alternative because they can provide higher bandwidth, lower interference, longer reach, and lower energy cost per transmitted bit.

As AI infrastructure expands, photonic communication is increasingly viewed as an important scaling option for some high-bandwidth systems. Beyond data movement, researchers are pursuing something even more ambitious: photonic computing itself.

This goes beyond using light simply to move information. It involves using optical systems to perform computation directly. However, it is worth drawing a clear distinction here, because these two ideas — photonic communication and photonic computing — are often conflated yet differ significantly in maturity and application. Photonic interconnects, which move data between chips, processors, or systems using light, are already commercially deployed in networking and are increasingly being pushed closer to compute packages and AI infrastructure. Photonic computing, by contrast, uses optical systems to perform mathematical operations — leveraging light waves, interference patterns, resonators, and waveguide networks to carry out certain calculations without converting back to electronics. This is especially intriguing for workloads involving linear algebra, signal processing, and machine learning, where some calculations may map efficiently to optical behavior.

The distinction matters because the engineering challenges, timelines, and commercial readiness of these two directions are fundamentally different. Photonic interconnects are a present-day industrial story. Photonic computing is a compelling but longer-horizon scientific pursuit.

The appeal is obvious. Photonic computing is attractive because it offers extremely high speed, natural parallelism, very low heat generation compared with conventional current-heavy systems, and potentially lower energy consumption for certain operations. Because light can propagate and interfere in ways that are highly useful mathematically, optical systems can perform some classes of computation very efficiently.

This field presents significant potential, although substantial engineering challenges persist. Fully photonic general-purpose computing remains a long-term objective rather than a current commercial reality. The active pursuit of this goal by researchers underscores the increasing limitations of conventional computing architectures. As the costs associated with data movement and processing escalate, concepts that were once considered unconventional are now being seriously evaluated for practical implementation.

In the near term, a hybrid approach is the most probable direction for computing system development.

Computing systems are unlikely to become solely optical anytime soon. Electronics remain highly effective for logic, control, memory management, and broad system functions. Meanwhile, photonics excels in communication and specific computational tasks. The most practical and powerful approach is a hybrid system that combines both. In these systems, electronic circuits perform what they do best, while photonic circuits take on ultra-fast data transfer and possibly other high-throughput tasks.

This hybrid approach aligns with historical trends in technology. The industry is moving toward Co-Packaged Optics (CPO). Players such as Broadcom were publicly presenting CPO as moving from demonstration toward volume production and broader deployment. In this architectural paradigm, optical transceivers are embedded inside the same advanced package as the GPU or ASIC, rather than on the edge of the circuit board. By reducing the distance electrons travel over lossy copper traces to just a few millimeters, CPO directly addresses the data movement bottleneck, significantly lowering power consumption and latency.

New paradigms seldom replace existing ones abruptly; instead, they typically complement established technologies, initially dominating areas where they offer the greatest benefits and gradually influencing the broader system. Photonics is expected to follow this trajectory, first excelling in applications where bandwidth, distance, and efficiency are critical, and subsequently expanding as manufacturing and design techniques advance.

Of course, there are serious obstacles.

Photonic chips are not easy to build. One of the most fundamental obstacles stems from silicon's material properties. Unlike direct-bandgap semiconductors such as indium phosphide or gallium arsenide, silicon has an indirect bandgap, which makes efficient light emission intrinsically difficult. Electrons in silicon lose energy as heat rather than by emitting photons, making it a poor candidate for on-chip laser sources. This is not simply a manufacturing issue; it arises from silicon's underlying band structure and therefore usually requires hybrid or external-light-source approaches. As a result, engineers face a persistent dilemma: either couple external lasers onto the chip, which introduces alignment complexity, coupling losses, and packaging overhead, or use hybrid integration approaches that bond light-emitting III-V semiconductor materials to silicon substrates. Both approaches add cost and complexity, and neither is as elegant as a fully monolithic solution. Until a practical path to efficient on-chip light generation in silicon is established, laser integration is likely to remain one of the major bottlenecks in silicon photonics.

Beyond this, managing optical losses, fabrication precision, and component alignment is demanding. Many optical elements are physically larger than electronic transistors, which complicates dense integration. In a world where billions of transistors can fit onto a fingernail-sized chip, optical components can feel frustratingly bulky. Engineers are working to shrink them while maintaining efficiency and performance, but the balance is delicate.

Manufacturing precision is another issue. Optical systems are less forgiving of misalignment than many electrical ones. Small errors in positioning or fabrication can affect signal quality, coupling efficiency, and system behavior. Heat management also does not disappear entirely just because photons are involved. Some optical components still generate heat or respond sensitively to temperature changes, so thermal control remains important.

Despite these challenges, the range of potential applications is broad and growing in significance.

AI supercomputing represents a primary application area. Large-scale AI systems can benefit significantly from photonic networks that transfer data between accelerators with reduced overhead. More broadly, supercomputing relies on the efficient coordination of thousands of processing elements, and optical interconnects may alleviate communication bottlenecks. The telecommunications sector, already closely linked to optical transport, is expected to further integrate advanced optical-chip functionality as bandwidth requirements increase.

This extends well beyond data centers. Undersea cable systems, which carry most of the international internet traffic, depend on optical amplification and coherent transmission technology to span thousands of kilometers with minimal signal degradation. Metropolitan networks connecting cities and regions, and the emerging fronthaul infrastructure linking 5G and future 6G base stations to their core networks, all represent expanding photonics markets where higher bandwidth, lower latency, and greater energy efficiency are persistent demands. In each of these environments, advances in photonic chip technology translate directly into greater network capacity and reduced operational costs.

Quantum communication is widely expected to rely heavily on photonic systems, since photons serve as effective carriers of quantum information over long distances. Additionally, photonics underpins next-generation LiDAR systems in autonomous vehicles, facilitating the shift from bulky mechanical scanners to compact, high-performance solid-state systems.

Beyond transportation and computing, photonics is making quiet but significant inroads into life sciences and healthcare. Optical biosensors built on photonic chip platforms can detect biological molecules with extraordinary sensitivity, enabling faster and more precise diagnostic tools. Photonic spectroscopy systems, which identify materials and compounds by analyzing how they interact with light, are finding applications in everything from point-of-care medical diagnostics to pharmaceutical quality control.

Medical imaging modalities such as optical coherence tomography, which produces high-resolution cross-sectional images of biological tissue, are already clinically deployed and stand to benefit substantially from further photonic integration. These applications share a common theme: the ability of light to interact with matter in highly specific and sensitive ways makes photonics a natural fit for measurement and detection tasks that electronics alone cannot perform as effectively.

To address these growing demands, the industry is investigating alternative materials beyond silicon to enhance performance, including Thin-Film Lithium Niobate (TFLN), also referred to as Lithium Niobate on

Insulator (LNOI), which is likely to enable superior high-speed modulation for next-generation 200G and 400G communication lanes.

Although fully optical computers may remain a long-term objective, photonic technologies are already influencing the structure of contemporary computing infrastructure. These technologies are being integrated into current systems, initially in communication-intensive roles and, progressively, into more fundamental architectural capacities.

This transition is therefore of considerable significance. The semiconductor revolution established computation based on electrons, while the next phase is likely to involve a division of functions between electrons and photons. Electronics will remain essential for logic, control, and dense switching, whereas photonics will increasingly address data movement, bandwidth, and specialized computational acceleration. The combination of these technologies may yield systems with capabilities surpassing those achievable by either approach independently.

The emergence of photonic chips represents more than a specialized advancement in optical engineering; it is integral to a broader reconsideration of computing paradigms in an era characterized by artificial intelligence, cloud infrastructure, and substantial data flows. The traditional model, which relied solely on increasing transistor speeds to drive progress, is now insufficient. Contemporary performance depends equally on the efficiency of information transfer and processing.

Moreover, light enables exceptionally efficient data transmission.

Photonic chips provide a pathway to faster data transmission, reduced power consumption, and more scalable architectures for future computing platforms. While they are unlikely to replace electronic chips completely, their significance may lie in enabling the continued scalability of modern computing systems by addressing data movement challenges.

In the coming decades, the most advanced computing systems are likely to depend on a synergistic partnership between electronic and photonic technologies.

Electronic circuits will continue to perform computational tasks, while photonic circuits may increasingly facilitate data communication.

The industry is now confronting challenges such as heat generation, bandwidth limitations, and data congestion in AI systems, leading to growing interest in photonic communication as a practical solution in selected bandwidth environments.

References

1. Intel, Intel Silicon Photonics.
2. Intel Research, Integrated Photonics: Transitioning to End-to-End Optical I O.
3. Intel Newsroom, Intel Demonstrates First Fully Integrated Optical I O Chiplet.
4. Intel, The 50G Silicon Photonics Link.
5. IBM Research, Photonic Computing.
6. IBM Research, Silicon Photonics Packaging.
7. IBM Research, III-V on Silicon Photonics.
8. IBM Research, Silicon Photonics Integration Technologies for Future Computing Systems.

9. IBM Research, A Monolithically Integrated Silicon Photonics 8×8 Switch in 90nm SOI CMOS.
10. IBM Research, Silicon Photonic Switching: Technology and Architecture.
11. IBM Newsroom, IBM Brings the Speed of Light to the Generative AI Era with Optics Breakthrough.
12. Broadcom, 2024, Broadcom Delivers Industry's First 51.2-Tbps Co-Packaged Optics Ethernet Switch Platform for Scalable AI Systems.
13. Journal of Optical Communications and Networking, 2024, Progress in Research on Co-Packaged Optics.
14. Nature Photonics OFC Technical Proceedings, 2026, Breaking the 1.6 Tb Barrier: Device Interoperability and 224G 448G PAM4 Signaling.

PART V: The Semiconductor Future

Chapter 18: The Rise of India as a Semiconductor Nation – Strategy, Scale and Self-Reliance

The future of the semiconductor industry will be influenced not only by established leaders but also by nations striving to secure a lasting position within the sector. Among these emerging participants, India has become a particular focus.

For decades, India's contribution to the semiconductor sector was substantial yet limited. The country supplied engineering talent, design expertise, embedded systems knowledge, research support, and software integration to the global chip industry. However, the physical aspects of semiconductor manufacturing largely occurred outside India. This imbalance characterized India's position in the semiconductor era for many years. While this history explains India's past role, it does not necessarily dictate its future trajectory.

India's significance in the future of semiconductors does not stem from having already attained the status of a leading semiconductor power. India has not yet matched Taiwan in advanced logic, South Korea in memory, the United States in leading-edge design and semiconductor tools, or Europe in lithography and specialized research and equipment. Assertions to the contrary conflate aspiration with achievement. The more pertinent question is whether India is beginning to develop the capabilities necessary to play a meaningful role in the next phase of the semiconductor era.

Evidence increasingly suggests that India is beginning to assemble these capabilities.

A clear indicator of progress is that India's semiconductor initiatives have moved beyond policy statements, design centers, and strategic declarations. These efforts are now materializing as infrastructure, partnerships, approved projects, pilot lines, workforce development programs, manufacturing facilities, and initial commercial output. This transition is significant because the future of semiconductors depends on tangible assets such as facilities, tools, talent pipelines, supplier networks, process expertise, and operational execution under demanding industrial conditions. A semiconductor ecosystem gains credibility only when it advances beyond policy intentions and becomes materially evident.

India has now reached this stage of tangible ecosystem development.

By early 2026, India's semiconductor program had expanded into a broad network of approved projects spanning multiple states and technology segments. This development is historically significant. For a nation previously recognized for its chip design talent but lacking manufacturing presence, the establishment of a physical semiconductor infrastructure represents a structural transformation. Discussions about India's semiconductor future now increasingly focus on current developments rather than solely on future aspirations.

India's emergence in the semiconductor sector is unlikely to commence at the industry's technological frontier. Instead, progress is expected through the development of layered capabilities, including assembly, testing, packaging, compound semiconductors, automotive and power electronics, design services, specialized devices, and selected manufacturing segments. This gradual approach reflects the typical evolution of robust semiconductor ecosystems. Nations achieve semiconductor prominence by establishing relevance across multiple layers of the value chain, allowing capabilities to accumulate over time.

At this juncture, India's semiconductor development trajectory is particularly noteworthy.

A prominent indicator of India's advancing semiconductor sector is the transition of major projects from policy proposals to industrial implementation. The significance of Micron's facility in Sanand extends beyond the establishment of a new plant; it symbolizes India's integration into the evolving global semiconductor landscape. Backend manufacturing, once considered peripheral, has become central in an era defined by chiplets, heterogeneous integration, and system-level performance. Assembly, testing, and packaging are now integral to semiconductor competitiveness. Micron's investment is therefore significant not only for attracting a global firm, but also for positioning India within a critical and increasingly valuable segment of the semiconductor value chain. Having commenced commercial operations in early 2026, the facility has begun exporting memory modules manufactured in India to global markets, demonstrating that India-based output meets the rigorous qualification standards of global OEMs.

Around the same time, Kaynes Semicon's OSAT facility marked India's entry into the vital field of power electronics, specializing in smart power modules for electric vehicles and renewable energy systems. This highlights the significance of Kaynes Semicon's growth, extending beyond a single plant. The key is in replicating processes across multiple projects. A semiconductor ecosystem becomes more credible when several facilities advance through stages such as hiring, qualification, disciplined production, and supplier partnerships. Each new site enhances manufacturing skills, workforce development, testing, identifying bottlenecks, building vendor relations, and accumulating industrial experience. Therefore, Kaynes Semicon stands as both a corporate milestone and proof that India is developing the operational stability necessary for sustained semiconductor growth.

For similar reasons, Tata Electronics, meanwhile, emerged as a foundational industrial anchor, with the Dholera fab project advancing and the Assam ATMP facility expanding India's capabilities in semiconductor assembly and manufacturing support. Together, these developments suggest that India's rise is no longer defined only by announcements and incentives, but by the gradual conversion of policy into production. The company's significance stems from its broad engagement across the emerging semiconductor ecosystem, positioning it as a key actor supporting both front-end and back-end capabilities. This comprehensive involvement is critical, as semiconductor ecosystems typically require a few institutions to operate across multiple value chain layers and maintain long-term commitments despite technical challenges and delayed returns.

The expansion of India's semiconductor ecosystem through projects such as HCL–Foxconn, SiCSem, 3D Glass Solutions, CDIL, and ASIP Technologies should be interpreted as more than mere geographic or numerical growth. The critical factor is the type of semiconductor future these investments are shaping. A more distributed, resilience-oriented global semiconductor landscape will favor countries offering diverse capabilities. Products such as display driver chips, silicon carbide devices, advanced packaging, glass substrates, interposers, and specialty semiconductor components exemplify a future where ecosystem diversity, rather than reliance on a single manufacturing segment, determines relevance. Consequently, India's value may arise from fostering the coexistence and mutual reinforcement of multiple strategically significant semiconductor layers.

This broader strategic direction closely aligns with the semiconductor industry's evolving nature. The sector is moving beyond the traditional focus on transistor miniaturization. Advanced packaging, heterogeneous integration, chiplets, system-level optimization, and innovative substrate technologies are increasingly central to performance, particularly in fields such as artificial intelligence, high-performance computing, data center architectures, and advanced electronics. In this context, a country does not need to immediately dominate the leading-edge logic node to achieve strategic relevance. Instead, value can be established by developing strengths in enabling layers of the technology stack, especially those involving integration, packaging, and materials.

India's development of assembly and test facilities, OSAT (Outsourced Semiconductor Assembly and Test) initiatives, compound semiconductor projects, and packaging-related investments should therefore be regarded as foundational elements for establishing a broader role in the global semiconductor industry.

A similar perspective applies to power semiconductors and specialized electronics. The future of semiconductor applications will extend beyond smartphones and cloud data centers to include electric vehicles, renewable energy systems, industrial automation, grid modernization, telecommunications hardware, defense systems, medical devices, and intelligent infrastructure. These sectors are highly dependent on semiconductors, though not necessarily on the most advanced logic nodes. This characteristic makes them particularly relevant for countries developing capabilities incrementally. India's advancements in automotive electronics, power systems, industrial modernization, and electronics manufacturing provide domestic impetus to engage in these semiconductor segments. Thus, India's semiconductor trajectory is connected not only to export ambitions but also to the broader transformation of its domestic economy.

The future relevance of India in the semiconductor sector will also depend on talent, which remains one of the country's most significant long-term strengths.

Global semiconductor companies have long depended on Indian engineers for design, validation, verification, firmware development, embedded systems, and software integration. This legacy is significant, as the development of a robust semiconductor sector requires skilled professionals with expertise in complex systems. While infrastructure, tools, and capital can be acquired relatively quickly, talent ecosystems must be cultivated over extended periods. Countries that achieve sustained success in semiconductors are those capable of consistently producing engineers, technicians, materials scientists, process specialists, equipment experts, and manufacturing leaders over multiple decades.

Consequently, workforce development is likely to be as critical as establishing manufacturing facilities. Programs such as Chips to Startups are helping expand the engineering pipeline needed to support design, packaging, testing, and, eventually, deeper manufacturing capability.

If India successfully integrates its extensive engineering talent with fabrication, packaging, testing, process control, and ecosystem-level execution, it will acquire one of the most valuable assets for a semiconductor nation: renewable technical capacity. While individual fabrication plants, firms, or projects may encounter setbacks, a robust technical base enables the regeneration of capabilities over time. Ultimately, this distinction may prove more consequential than any single project announcement.

A similar analytical approach applies to India's efforts in semiconductor design, including the expansion of advanced work through Renesas, the broader promise of indigenous processors represented by DHRUV64, and the strategic significance of VIKRAM3201, qualified for use in critical sectors such as space exploration and advanced telecommunications infrastructure. These developments are important not only as indicators of technical progress but also as signals of the type of semiconductor nation India could realistically become. India's future strength is unlikely to result from replicating the historical trajectories of established semiconductor powers. Instead, it is more likely to emerge from integrating design expertise, strategic chip capabilities, selective manufacturing scale, advanced packaging, and ecosystem diversity into a model tailored for a fragmented, security-focused global environment. In this context, the convergence of trusted domestic chips for strategic sectors, enhanced design leadership, and increased manufacturing depth may collectively define India's most credible path forward in the semiconductor industry.

The connection between indigenous expertise and strategic capability is also evident in institutions established prior to the recent wave of semiconductor expansion. The modernization of the Semiconductor Laboratory (SCL) in Mohali, supported by substantial government investment, is significant for reasons extending beyond the facility itself. SCL embodies continuity, linking earlier phases of India's semiconductor history to current developments. If effectively upgraded, SCL can function not only as a fabrication asset but also as a bridge connecting public-sector technical capacity, prototyping, research, student-designed chips, and national mission objectives. Thus, the foundations of India's semiconductor future will derive not only from new private projects but also from the revitalization of established institutions.

This broader perspective is essential because the development of a semiconductor nation depends not solely on the establishment of new factories but on the creation of comprehensive ecosystems.

One of the most significant recent developments is the expansion of India's semiconductor strategy beyond fabrication and packaging plants to encompass the broader ecosystem, including raw materials, specialty inputs, substrates, process technologies, design intellectual property, and equipment. This shift is critical, as semiconductor plants rely on a complex network of chemicals, gases, specialty materials, testing systems, clean-room support, engineering services, and equipment-related capabilities. If these supporting layers remain external, operational facilities will continue to operate within a fragile, dependent structure.

Therefore, a genuine semiconductor nation requires industrial depth rather than mere industrial visibility. This underscores the importance of expanding the semiconductor ecosystem under India Semiconductor Mission 2.0. Initiatives to promote domestic production of semiconductor equipment and materials, specialty chemicals, and high-purity gases, strengthen supply chains, and enhance full-stack semiconductor capabilities represent a more advanced stage of policy development. This approach acknowledges that lasting national relevance depends on projects reinforcing each other through a robust industrial base, rather than relying on temporary momentum.

While the accumulation of projects may attract attention, the formation of a comprehensive ecosystem is what establishes enduring national capability. The distinction between these approaches is between temporary industrial enthusiasm and a sustainable position in the future global semiconductor landscape. The objective is clear: facilities in places such as Dholera and Sanand must evolve beyond isolated industrial sites and become part of a broader, more self-sustaining supply chain.

The increasing prominence of SEMICON India holds significance that extends beyond the event itself. By consistently hosting the event at scale, India enhances its credibility within the global semiconductor industry. In a sector where trust is established through sustained engagement, ecosystem maturity, and demonstrable commitment, such a platform is essential. It demonstrates that India regards semiconductors not as a temporary policy focus, but as a long-term national priority capable of attracting the world's leading firms, suppliers, researchers, policymakers, and investors.

Geopolitical timing further enhances India's relevance in the future semiconductor landscape. In an era increasingly defined by supply chain insecurity, technological rivalry, and strategic realignment, India's value lies not only in its market size and engineering talent but also in its emergence as a – trusted geography. Its semiconductor growth has been reinforced by closer strategic coordination with the United States, Japan, and the European Union, including the India-U.S. TRUST initiative and the concluded India-EU Free Trade Agreement (FTA). These deep diplomatic alignments are more crucial because they help position India within the higher-value, more sensitive layers of the technology ecosystem. In this context, self-reliance does not mean isolation. It means becoming a secure, credible, and indispensable node within a broader network of technologically advanced partners. That may prove to be one of the most important foundations of India's semiconductor rise.

Global firms and governments are increasingly seeking to diversify supply chains, reduce geographic concentration, and enhance the resilience of digital infrastructure. While established semiconductor centers remain important, new locations are now being considered with greater seriousness. India is well-positioned to benefit from this shift due to its scale, strategic location, political significance, market demand, and an established technology workforce. In a more fragmented, security-focused, and resilience-driven global environment, these attributes take on greater importance.

However, opportunity alone does not guarantee success.

India's prospects as a semiconductor nation will depend on its ability to address the complex challenges faced by advanced semiconductor ecosystems worldwide. These challenges include ensuring stable power, clean water, effective chemical handling, efficient logistics, contamination control, supplier depth, precision infrastructure, equipment maintenance, technical training, process discipline, reliable execution, and policy continuity. Semiconductor manufacturing is highly demanding and does not reward broad intentions if operational standards are not consistently met. Success in this industry requires the coordinated and sustained execution of numerous complex tasks over extended periods.

For these reasons, the concept of the "rise of India as a semiconductor nation" should be approached with both seriousness and caution.

Serious consideration is warranted because the ongoing transformation is substantive. India is evolving beyond its traditional roles as a design hub or consumer market, establishing a more tangible and diversified presence in semiconductors through manufacturing facilities, partnerships, strategic chip initiatives, skills programs, pilot lines, packaging capabilities, materials-related investments, institutional modernization, and ecosystem planning. This progression is historically significant in itself.

Caution is also necessary, as progress should not be conflated with completion. The history of the semiconductor industry includes numerous costly initiatives that failed to achieve lasting scale, with many abandoned plans, delayed milestones, execution gaps, and instances of strategic overconfidence. India is not immune to these risks. Achieving durable success will require sustained investment, institutional patience, coordination between central and state governments, global partnerships, supply chain development, and a commitment to continuous improvement through execution rather than premature declarations of success.

Nevertheless, despite these cautions, the overall trajectory remains significant and warrants attention.

India's potential future role in the semiconductor sector is no longer limited to a single area of expertise. The country may emerge as a nation that integrates design strength, packaging capability, compound semiconductor relevance, selected manufacturing segments, power and specialty device production, materials and equipment ambitions, talent scale, strategic chip capabilities, and geopolitical significance. While this combination may not position India as the sole center of the semiconductor world, it could establish the country as an indispensable node within the global network.

This may represent a more realistic definition of a semiconductor nation in the coming decades.

The future global semiconductor landscape is unlikely to be dominated by a single geography. Instead, it will be more distributed, security-oriented, ecosystem-driven, and shaped by resilience rather than solely by efficiency. In this context, countries capable of providing multiple forms of value will assume greater importance. India has the opportunity to become one of these key contributors.

India's ascent should not be viewed as a sudden transition from absence to dominance. Rather, it is more accurately characterized as a gradual yet significant progression from limited participation to broader relevance, and ultimately toward greater ecosystem depth.

This progression remains ongoing. It is not yet complete. Its outcome is not assured. However, it is no longer merely hypothetical.

India's future in semiconductors will be determined not by its ability to rapidly match established industry leaders, but by its capacity to steadily cultivate, deepen, and sustain semiconductor capabilities over time.

If this occurs, India will not merely participate as a consumer in the semiconductor future. India could become a more consequential contributor to the future semiconductor industry.

References

1. India Semiconductor Mission, Official Website.
2. MeitY and Press Information Bureau, Under Chips to Startups.
3. Press Information Bureau, Bolstering India's Semiconductor Ecosystem.
4. Press Information Bureau, India Semiconductor Mission 2.0.
5. Press Information Bureau, 2024, Cabinet Approves One More Semiconductor Unit under India Semiconductor Mission.
6. Prime Minister's Office (PMO) India / PIB, 2026, PM inaugurates Kaynes Semicon Plant at Sanand, Gujarat.

7. Reuters, 2026, India Seeks Dutch Knowhow in Semiconductor Push.
8. Reuters, 2024, India Gives Green Light to Chip Plants Worth $15.2 Billion.
9. Micron, 2026, Micron Celebrates Opening of India's First Semiconductor Assembly and Test Facility.
10. Tata Electronics, Tata Group to Build the Nation's First Fab in Dholera.
11. Press Information Bureau, 2026, Assam Semiconductor Facility Note.
12. CG Power, Renesas, and Stars Microelectronics, Joint Announcement.
13. CG Semi, Official Website.
14. Times of India, 2026, CG Semi Set to Operationalize OSAT Soon.
15. White House iCET Fact Sheet, United States and India Elevate Strategic Partnership.
16. European Commission Press Corner
17. U.S. Department of State
18. Qualcomm, Design in India Program.
19. Texas Instruments, TI India Fact Sheet.
20. Tata Electronics and Qualcomm, 2026, Partner to Manufacture Qualcomm Automotive Modules in India.
21. Tata Electronics and Intel, 2025, Strategic Alliance to Establish Silicon and Compute Ecosystem in India.
22. Tata Electronics and Synopsys, Collaboration for India's First Fab.

Chapter 19: The Trillion-Dollar Chip Economy – Value Beyond the Visible Market

Semiconductors have evolved beyond their traditional roles as a strategic sector, manufacturing domain, or technical specialty confined to engineering circles. They form a foundational enabling layer of the modern economy, deeply embedded in production, infrastructure, mobility, communications, finance, defense, and computation. The performance and continuity of many industries increasingly depend on the chip ecosystem.

The term “trillion-dollar chip economy” signifies more than projected revenue growth; it reflects a profound structural transformation. Semiconductors are transitioning from important product components to the foundational layer supporting extensive segments of economic activity. Modern smartphones, automobiles, data centers, and factories all rely on intricate combinations of processors, memory, power devices, sensors, controllers, connectivity chips, and advanced packaging. The semiconductor industry no longer simply supplies products; it increasingly enables the systems that drive the contemporary economy.

Semiconductors, therefore, occupy a distinctive position within the modern global economy. While their direct market size is substantial, their indirect economic influence is even greater. Although a chip may represent only a small portion of a product’s total cost, its absence can halt production entirely. This distinguishes semiconductors from many other industrial goods. Their significance derives not only from their individual value but from the capabilities they enable. For example, a missing processor can delay server deployment, a missing microcontroller can halt automobile assembly, a missing power semiconductor can impede electric vehicle platforms, and a missing radio-frequency chip can undermine communications infrastructure. These small components now have system-level economic implications.

A significant yet less visible shift is that the semiconductor industry can no longer be evaluated solely by the value of chips sold in the open market. Increasingly, semiconductor value is embedded in custom silicon designed by device manufacturers, proprietary chips developed by hyperscale companies, and the growing importance of advanced packaging and system integration. The chip economy increasingly extends beyond traditional sales-based measurement boundaries. Its actual scale is broader, more complex, and more deeply integrated into modern technology than conventional revenue figures indicate.

This transformation affects not only how the industry is measured, but also how it is controlled. When leading technology companies design chips tailored to their specific workloads, data centers, devices, and software ecosystems, they capture a greater share of the value chain. The balance of power is shifting. Semiconductor leadership is increasingly shaped not only by firms that manufacture or sell standard chips, but also by those that design architectures, define workloads, own platforms, and integrate hardware with software at scale. Consequently, the chip economy is becoming increasingly strategic as it expands.

The contemporary semiconductor economy can be best understood as a multiplier economy. Improvements in chip performance, energy efficiency, memory bandwidth, integration, or packaging capability have cascading effects across multiple sectors. Enhanced chips lead to superior cloud infrastructure, which in turn enables more advanced artificial intelligence. Improved AI transforms software, manufacturing, healthcare, logistics, and enterprise decision-making. More efficient power semiconductors advance electric mobility, industrial drives, renewable energy conversion, and charging systems. Advanced sensors and edge processors bolster automation, robotics, and smart devices. Although semiconductors may appear to be a specialized upstream industry, they are among the most significant amplifiers of downstream productivity in the global economy.

This multiplier effect accounts for the intense investment attracted by the semiconductor sector. Few industries demand such extensive and sustained commitments of capital, talent, research, and organizational discipline. The costs associated with constructing and equipping advanced fabrication facilities have reached unprecedented levels. Beyond fabrication, the ecosystem requires ongoing investment in design tools, intellectual property, materials science, process technology, lithography, packaging, testing, validation, and software integration. Companies and governments persist in investing because the benefits extend well beyond chip sales, conferring strategic leverage across the digital economy to those who advance semiconductor capabilities.

The expansion of the semiconductor industry is driven by diverse and simultaneous sources of demand. Artificial intelligence is the most prominent, requiring accelerated computing, high-bandwidth memory, advanced interconnects, and energy-efficient data center architectures. However, AI represents only one aspect of a broader landscape. Cloud computing is expanding, electric vehicles now incorporate significantly more semiconductor content, telecommunications networks are becoming denser and more software-defined, and industrial systems are increasingly automated and sensor-rich. Consumer electronics continue to advance, and the Internet of Things extends computation into everyday environments. The critical factor is not any single trend, but the convergence of multiple trends that collectively intensify the demand for more advanced and specialized chips.

Equally significant is the uneven growth across segments within the expanding chip economy. The large share of current growth and value creation is concentrated in artificial intelligence-related demand, leading-edge logic, and high-bandwidth memory, while mature and less differentiated segments experience slower growth. This dynamic results in an industry that is not only larger but also more concentrated, with competitive advantage increasingly accruing to those who control the most advanced aspects of design, manufacturing, packaging, memory architecture, and system integration.

This uneven distribution is significant because semiconductor growth is rarely uniform across the value chain. Certain segments expand due to frontier demand, while others encounter pricing pressures, commoditization, or cyclical slowdowns. Although the industry is approaching a trillion-dollar scale, profits are not distributed evenly. Instead, they concentrate around advanced control points such as leading-edge design, hyperscale computing platforms, top-tier foundry capacity, advanced memory, critical equipment, core software tools, and sophisticated packaging. As the industry grows, this concentration may intensify rather than diminish.

This shift is also altering the composition of value within the semiconductor industry. Growth is increasingly driven by complexity, specialization, and system integration rather than by higher volumes of standardized devices. Advanced logic, AI accelerators, high-bandwidth memory, chiplets, heterogeneous integration, and sophisticated packaging are generating new concentrations of value. While earlier progress was primarily measured by transistor scaling, economic power now accrues across a broader range of capabilities, including architecture, packaging, interconnects, power efficiency, software ecosystems, manufacturing discipline, and access to critical tools and materials. The future value of the industry will be determined not only by producing more silicon, but by developing more capable silicon systems.

A comprehensive understanding of the chip economy requires recognition of its layered business structure. Value is distributed among fabless designers, integrated device manufacturers, foundries, memory producers, equipment makers, design software firms, materials suppliers, outsourced assembly and test specialists, and increasingly, advanced packaging companies that integrate multiple dies into functional systems. Some firms define architecture, others provide manufacturing capacity, some control essential tooling, and others dominate memory or specialize in packaging, integration, testing, or specialty materials. The semiconductor economy comprises an interconnected array of businesses, each characterized by distinct economics, risks, and strategic significance.

This complexity underscores why the semiconductor economy cannot be evaluated solely through revenue figures. It is also defined by intricate dependencies. A small number of firms occupy critical chokepoints in areas such as design automation, lithography, foundry manufacturing, memory, deposition and etch, materials, and assembly and packaging. The global chip economy is thus both expansive and concentrated: it operates across continents but relies on a limited set of highly specialized competencies that are challenging to replicate. This combination of global reach and concentrated expertise imparts semiconductors with exceptional economic significance. Their impact is widespread, yet their capabilities are closely held.

Simultaneously, the semiconductor industry does not experience linear growth. Its history is marked by cycles, including periods of shortage followed by overcapacity, surges in demand followed by inventory corrections, and phases of exuberant investment followed by slower absorption. Memory markets are particularly prone to such fluctuations, but cyclical pressures affect the entire ecosystem in various ways. Even as the industry approaches a trillion-dollar scale, it remains subject to cyclical dynamics. The current era is marked not by the absence of volatility, but by the growing structural importance of semiconductors despite ongoing fluctuations.

This distinction is significant. While cyclicality influences pricing, margins, and capital expenditures in the short term, the long-term significance of semiconductors in the global economy continues to grow. Although the industry may experience temporary corrections, each phase of digitization increases societal dependence on chips. Following every downturn, semiconductors reemerge not as a transient technology, but as an increasingly essential foundation for computing, mobility, connectivity, automation, and intelligence. Market volatility is transient, whereas technological centrality endures.

As a result, semiconductor capabilities now influence the balance of industrial power beyond the confines of the chip sector. Nations compete for advanced fabrication facilities not only for employment or prestige, but because semiconductor capabilities affect resilience, innovation capacity, export competitiveness, digital infrastructure, and strategic autonomy. Companies vie not only to sell processors or memory, but also to define the computational foundations of future markets. Thus, semiconductor competition extends beyond market share within a single industry; it may influence which entities control some of the critical technological infrastructure in the next industrial era.

For these reasons, the emergence of a trillion-dollar semiconductor economy represents more than a milestone in sectoral expansion. It signifies the elevation of semiconductors to a core enabling layer of the twenty-first-century economy. Semiconductors now occupy a central position in industrial development, determining the performance of machines, the intelligence of networks, the efficiency of factories, the capabilities of vehicles, the scale of data centers, and the pace of technological advancement. Whereas chips were once regarded merely as inputs to growth, they are now recognized as fundamental structures shaping the organization of economic progress.

Semiconductors originated as components within electronic systems but have evolved into the core technological substrate underpinning much of modern economic life. Their scale, influence, and strategic significance continue to grow. The rise of a trillion-dollar chip economy is not merely the narrative of a single industry's prosperity; it is the story of silicon becoming increasingly important to the future distribution of global economic and technological power.

References

1. McKinsey & Company, The Semiconductor Decade: A Trillion-Dollar Industry.
2. McKinsey & Company, 2026, Hiding in Plain Sight: The Underestimated Size of the Semiconductor Industry.
3. McKinsey & Company, 2024, McKinsey on Semiconductors 2024.
4. McKinsey & Company, Exploring New Regions: The Greenfield Opportunity in Semiconductors.
5. McKinsey & Company, Semiconductors Have a Big Opportunity, but Barriers to Scale Remain.
6. World Semiconductor Trade Statistics, 2025, Global Semiconductor Market Grows 26% in 2025 to $795.6 Billion.
7. World Semiconductor Trade Statistics, 2025, Global Semiconductor Market Show Continued Growth in Q2 2025.
8. World Semiconductor Trade Statistics, 2024, Semiconductor Market Forecast Spring 2024.
9. World Semiconductor Trade Statistics, Historical Billings Report.

10. Semiconductor Industry Association, 2024, State of the U.S. Semiconductor Industry.
11. Semiconductor Industry Association, 2025, State of the U.S. Semiconductor Industry.
12. Semiconductor Industry Association, Essential Innovation for U.S. Technology Leadership.
13. Semiconductor Industry Association, 2024, Databook.

Chapter 20: The Next 50 Years of Silicon – Reinvention at the Edge of Limits

For over fifty years, semiconductors have achieved a rare feat in industrial history by improving rapidly, consistently, and extensively enough to reshape civilization. From the advent of integrated circuits to the processors powering artificial intelligence, cloud computing, smartphones, satellites, and medical systems, the semiconductor industry has served as a primary driver of modern technological progress. While Moore's Law provided a guiding framework, the broader reality was that the industry continually discovered methods to reduce the cost of processing, transmitting, storing, and utilizing information. This transformation has fundamentally altered society.

The coming decades will differ significantly from the previous era. The traditional approach of shrinking transistors is encountering increasingly difficult physical and economic barriers. Issues such as leakage, thermal density, interconnect delay, manufacturing complexity, and the escalating costs of advanced fabrication facilities are constraining the classical trajectory. However, the evolution of computing will not cease with the end of straightforward scaling. Instead, it will become more diverse, architectural, and system-oriented, relying on the integration of multiple technologies rather than focusing solely on miniaturization. The future of silicon will be defined by an expanding set of tools and approaches.

This expanding toolkit is already evident in current research. Silicon is now advancing not only through further transistor miniaturization but also through the incorporation of new materials, device geometries, and manufacturing strategies. Atomically thin semiconductors such as MoS_2 and InSe, as well as novel ultrahigh-κ dielectrics, are being investigated for future low-power transistor concepts to maintain electrostatic control and reduce switching energy at scales, where traditional structures face limitations. Concurrently, the development roadmap for core logic is progressing toward gate-all-around devices, complementary FETs, and ultimately more advanced three-dimensional transistor configurations. The focus of innovation has shifted from merely reducing transistor size to the intelligent redesign of the entire device stack.

This transition is significant because it redefines the concept of progress in computing. Historically, advancements in computing were primarily associated with smaller transistors. In the near future, improvements are expected from coordinated developments across transistor architectures, interconnects, packaging, photonics, specialized accelerators, and software-assisted design. Interconnects are emerging as major constraints on performance and energy efficiency, indicating that future advancements will depend as much on component communication as on the miniaturization of switching elements. Consequently, progress will be characterized by the intelligent orchestration of entire systems rather than reliance on individual transistor innovations.

This orchestration will increasingly occur in three dimensions. Research into CFETs, stepped-channel FETs, monolithic three-dimensional integration, and advanced dense packaging indicates that future performance gains will derive as much from vertical stacking as from further planar scaling. Logic components are likely to move closer to memory - specifically through the integration of advanced generation HBM4 (High Bandwidth Memory) and beyond with tighter package integration, reducing latency and increasing usable bandwidth. Packaging is likely to transition from a secondary consideration to a central aspect of design. As a result, future chips may resemble layered computing structures in which power, memory, logic, and communication are optimized collectively.

In early 2026, backside power delivery (BSPDN) had begun to shift from the realm of experimentation to early commercial implementation at the leading edge. Intel, as the industry's first mover, has successfully integrated its PowerVia technology into high-volume manufacturing, while TSMC is positioning its A16 (1.6nm) node with Super Power Rail as a next-generation backside-power offering. Beyond these first implementations, BSPDN may become a more common architectural feature across leading-edge nodes into the 2030s, progressively converging with buried power rails and three-dimensional transistor integration to further decouple power delivery from signal routing at atomic scales. Future chips may resemble layered computing structures in which power, memory, logic, and communication are optimized collectively.

This context underscores the growing importance of chiplets. The future silicon platform is unlikely to consist of a single monolithic die addressing all computational needs. Instead, it will likely be a modular system composed of interoperable functional blocks, including CPU tiles, GPU tiles, AI accelerators, memory stacks, analog components, I/O dies, and domain-specific engines, all interconnected through advanced packaging and standardized interfaces. This approach enhances yield, promotes reuse, increases design flexibility, and enables cost amortization, particularly as leading-edge manufacturing becomes increasingly expensive. The dominance of more general-purpose processors across all workloads is increasingly being supplemented by a paradigm of federated computing blocks, enabling very high package power densities that are straining conventional cooling methods. This shift introduces a "thermal wall," necessitating a fundamental rethinking of the physical environment. The future of silicon will increasingly rely on liquid-based cooling methods, as traditional air cooling is supplanted by direct-to-chip liquid cooling and immersion systems. In this context, chip design must prioritize not only electrical efficiency but also optimal heat transfer to liquid media, establishing thermal management as a core architectural consideration.

Photonics will likely be one of the important pillars of that federated future. As systems scale, moving data efficiently becomes one of the biggest obstacles in computing. Optical I/O chiplets, co-packaged optics, and silicon photonics point toward a world in which future performance depends less on making one core slightly faster and more on moving colossal volumes of data with less heat and less power loss. In plain terms, the next age of silicon may need light as much as it needs electrons. If AI infrastructure continues to expand rapidly, the interconnect bottleneck is likely to become more acute, making photonics increasingly practical in selected high-bandwidth systems. The material landscape of silicon would also become increasingly hybrid. While silicon will remain the industrial foundation due to its extensive supporting ecosystem, it will increasingly be complemented by materials optimized for specific performance requirements.

Gallium nitride and silicon carbide are already gaining industrial significance in power electronics, radio frequency systems, and electrified mobility. Other materials, such as Ga_2O_3 and diamond, remain largely experimental but represent the frontier of potential semiconductor advancements. The so-called post-silicon era is unlikely to involve the obsolescence of silicon; rather, it will involve silicon being augmented by specialized materials.

Another significant development is the increasing role of artificial intelligence in hardware design for future computing systems. AI-assisted floor-planning and layout tools have demonstrated the ability to deliver results at speeds and scales beyond the capacity of human teams alone. Over time, this may transform chip design into a more automated and model-driven discipline. Advanced chips will most likely facilitate AI training, while AI will contribute to the design of even more advanced chips. This feedback loop may become a key driver of progress in the next semiconductor era, not by replacing engineers, but by expanding the range of design possibilities available to them. The next step may involve more autonomous AI-assisted design workflows, though that remains an emerging direction rather than an established industry standard.

Yet the hardware advances enabled by AI-assisted design will only be fully realized if the software stack keeps pace. Heterogeneous computing architectures — composed of CPUs, GPUs, AI accelerators, and domain-specific chiplets — pose a profound challenge for compilers, programming models, and operating systems. Today's software ecosystems were largely built around homogeneous, general-purpose processors; adapting them to efficiently orchestrate federated, heterogeneous systems remains an unsolved problem of considerable depth. Frameworks, abstraction layers, and compiler toolchains must evolve to automatically partition workloads, manage data movement across disaggregated memory hierarchies, and exploit the distinct capabilities of each compute element. Without this evolution, the performance potential of next-generation silicon will remain partially stranded — accessible to a narrow set of expert programmers but invisible to the broader ecosystem. Over the coming decades, the software-hardware co-design challenge may prove as consequential as the transistor roadmap itself.

The coming decades will most likely be characterized by the rise of heterogeneous computing. Future systems are expected to integrate CPUs, GPUs, AI accelerators, networking silicon, photonic interconnects, and, in certain applications, neuromorphic or quantum co-processors. Neuromorphic systems will be relevant for sparse, event-driven, ultra-low-power intelligence, while quantum systems will be advantageous in areas such as chemistry, optimization, materials discovery, and security.

One promising near-term path to scalable quantum hardware may involve silicon spin-qubit approaches that can leverage CMOS-adjacent fabrication expertise. Spin-qubit approaches, as pursued by Intel and other organizations, utilize existing CMOS fabrication infrastructure to construct qubits from individual electron spins in silicon quantum dots. This indicates that the relationship between quantum and classical silicon is not simply one of coexistence but of potential convergence. Quantum logic elements could eventually be integrated into or alongside conventional silicon dies, using the same fabrication processes as current processors. Regardless of whether this vision is realized in twenty or fifty years, it represents a significant long-term possibility for silicon's role in computing.

Neither technology is expected to replace classical semiconductors for general-purpose computing. Instead, both are likely to expand the broader computing landscape. Future computing architectures will likely consist of a federation of specialized computational units rather than a single dominant processor.

The development of this federation will extend beyond data centers and personal devices. A significant yet often underappreciated direction for silicon over the next fifty years is the proliferation of intelligence at the extreme edge, including industrial controllers, autonomous vehicles, agricultural sensors, infrastructure monitoring systems, and consumer wearables. These applications require a fundamentally different design philosophy compared to high-performance computing: ultra-low power consumption, extended operational lifetimes, robustness in harsh environments, and hardware-level security. Semiconductors designed for edge applications will be defined less by peak performance benchmarks and more by energy harvesting, continuous sensing, and the capacity to perform meaningful inference on minimal power. As embedded intelligence becomes pervasive across the physical world, the aggregate societal impact of edge silicon may ultimately rival that of data centers, even if individual devices remain less powerful.

The semiconductor manufacturing roadmap will become increasingly challenging. High-numerical-aperture extreme ultraviolet (High-NA EUV) lithography will extend patterning capabilities at the leading edge, while emerging techniques such as directed self-assembly and advanced deposition may support future generations if their defect and integration issues are addressed. However, each technological advance brings greater cost and complexity. As a result, the future of silicon is shaped not only by technical considerations but also by economic and geopolitical factors. The construction of leading-edge fabrication facilities requires substantial investment, packaging ecosystems are becoming more regionally anchored in some cases, and governments increasingly treat semiconductor capacity as strategic infrastructure. The industry's trajectory will depend on both technological feasibility and the resources nations and firms can commit.

This industrial dimension is critical. As the economics of transistor scaling become increasingly challenging, value creation is likely to shift toward packaging, integration, design automation, domain-specific architectures, and supply chain resilience. Regions unable to compete at the most advanced logic nodes may still play significant roles through outsourced semiconductor assembly and test (OSAT), assembly, test, marking, and packaging (ATMP), materials, specialty semiconductors, equipment ecosystems, and advanced packaging. Consequently, the global semiconductor landscape is expected to become more distributed, even if the most advanced manufacturing remains concentrated among a few entities. The future of silicon will be shaped not only in cleanrooms, but also through policy, capital allocation, standards development, and ecosystem-building.

The semiconductor roadmap for the coming decade is already taking shape. In the near term, technologies such as gate-all-around transistors, backside power delivery, chiplets, and advanced packaging will become increasingly central. Over the 2030s and beyond, complementary FETs, more advanced optical interconnects, monolithic three-dimensional integration, and broader heterogeneous integration are expected to define the high-performance frontier. Over the 2040s and beyond, computing platforms may become deeply layered, integrating logic, memory, sensing, communications, and specialized accelerators in configurations that surpass current chip categorizations. Silicon is likely to remain central, serving as the substrate and organizing framework for a more complex computing ecosystem.

At the furthest frontier of this expansion is the intersection of silicon with biology, a convergence that is reshaping both medicine and the fundamental architecture of information storage. Neural implants, bioelectronic medicine, and brain-computer interfaces require a distinct engineering philosophy compared to mainstream approaches. Success in these domains will be measured by biocompatibility, chronic tissue stability, and microwatt power budgets rather than processing speed.

Equally significant is the development of synthetic DNA storage, which encodes digital information into engineered nucleotide sequences at densities unattainable by silicon architectures, theoretically reaching exabytes per gram, stable for millennia, and requiring no active power. Importantly, DNA storage would not be a direct substitute for silicon computing and, in practice, would depend heavily on semiconductor-enabled instruments such as sequencers, sensors, and lab-on-chip systems. High-speed DNA writers, nanopore sequencing arrays, CMOS-integrated biosensors, and microfluidic lab-on-chip systems are all semiconductor technologies, and their advancement through silicon scaling will determine the pace at which molecular storage could expand beyond a niche archival to mainstream infrastructure. The probable long-term outcome is stratification rather than substitution, with silicon remaining the high-speed engine of active computation and synthetic DNA serving as an ultra-dense, effectively permanent cold storage layer for humanity's expanding archives. Over the next fifty years, silicon's influence on biological systems and molecular data infrastructure may rival or even redefine its transformative impact on communications and computation over the previous fifty years.

If these developments occur, the societal significance of semiconductor innovation will expand further. Future chips will be important not only for increasing device speed, but also for enabling climate modeling, accelerating drug discovery, optimizing energy grids, designing new materials, coordinating transportation, powering scientific instruments, and supporting distributed intelligence throughout the physical world. Semiconductors will thus serve not merely as products, but as essential tools for addressing large-scale societal challenges.

Despite the emergence of new materials, optical technologies, neuromorphic systems, quantum computing, and increasing architectural complexity, silicon is expected to remain indispensable. Its extensive manufacturing infrastructure, design tools, process expertise, and robust supply chain ensure its continued role as the industry's foundational material for decades to come. Many technologies considered as alternatives are, in practice, being developed using silicon-compatible processes. Silicon's enduring legacy may not be its solitary dominance, but rather its role as the platform enabling a diverse, multi-technology future.

The next fifty years of silicon will likely not follow a linear trajectory or replicate another era of straightforward Moore's Law scaling. Instead, this period will be characterized by layered progress, including new transistor geometries, novel materials, advanced packaging, optical communication, AI-assisted design, specialized accelerators, and an industry increasingly defined by the orchestration of diverse computational forms. While the future of civilization will continue to rely on silicon, it will do so as part of a broader, more integrated technological landscape.

References

1. Intel Foundry, Intel 18A.
2. Intel, Intel Demonstrates Breakthroughs in Next-Generation Transistor Scaling for Future Nodes.
3. Intel Newsroom, 2026, CES 2026: Intel Core Ultra Series 3 Debut as First Built on Intel 18A.
4. TSMC, 2026, A16 Technology.
5. IBM Research, Monolithic Stacked FET with Stepped Channels for Future Logic Technologies for IEDM 2024.
6. ASML, 5 Things You Should Know About High-NA EUV Lithography.
7. Nature, High-κ Perovskite Membranes as Insulators for Two-Dimensional Transistors.
8. Nature Materials, 12-inch Growth of Uniform MoS_2 Monolayer for Integrated Circuit Manufacture.

9. Nature Communications, Wafer-scale Functional Circuits Based on Two-Dimensional Semiconductors with Fabrication Optimized by Machine Learning.
10. imec, introducing 2D-Material Based Devices in the Logic Scaling Roadmap.
11. IEEE International Electron Devices Meeting, 2025, Proceedings.
12. Nature Reviews Electrical Engineering, Towards Fab-Compatible Two-Dimensional Electronics.
13. Nature Biotechnology, 2025, Scaling DNA Data Storage via CMOS-Integrated Electrochemical Synthesis.
14. IEEE Journal of Solid-State Circuits, 2026, Ultra-low-power Neuromorphic ICs for Chronic Neural Decoders.
15. INU Roadmap, 2026, Roadmap of Ultra-Thin Crystalline Silicon-based Bioelectronics.
16. Infineon Technologies, Infineon Pioneers World's First 300 mm Power Gallium Nitride Wafer Technology.
17. Power Electronics News, Wolfspeed Achieves 300-mm SiC Wafer Production.
18. imec, Record-Breaking RF GaN-on-Si Transistor Performance for High-Efficiency 6G Power Amplifiers.
19. MIT News, MIT Engineers Grow High-Rise 3D Chips.
20. Stanford Report, Researchers Unveil Groundbreaking 3D Chip to Accelerate AI.
21. Intel Newsroom, Intel Demonstrates First Fully Integrated Optical I/O Chiplet.
22. Electronics For You, World's First UCIe Optical Chiplet.
23. Nature Communications, Roadmapping the Next Generation of Silicon Photonics.
24. Nature Reviews Electrical Engineering, Future Interconnect Materials for Highly Integrated Semiconductor Devices.
25. Google DeepMind, How AlphaChip Transformed Computer Chip Design.

26. IBM, Quantum Roadmap.

AFTERWORD: The Invisible Power

The modern world is built on something we rarely see.

From a grain of sand to billions of transistors, semiconductors have become the quiet foundation of economies, technologies, and modern life. What was once a component has become a defining enabler of progress.

As the limits of traditional scaling come into view, the future will be shaped not only by smaller chips, but by smarter architectures, new materials, and bold ideas. Artificial intelligence, quantum computing, and other emerging technologies are already expanding the possibilities of what comes next.

The central question, then, is not only who will build the future of computation, but how that future will be shaped, shared, and sustained.

In that answer lies much of the technological and economic direction of the century ahead.

www.ingramcontent.com/pod-product-compliance
Lightning Source LLC
Chambersburg PA
CBHW031322230926
51848CB00001B/58
* 9 7 8 8 1 9 3 8 8 9 5 3 4 *